# HOMOGENEOUS BANACH ALGEBRAS

# PURE AND APPLIED MATHEMATICS

*A Program of Monographs, Textbooks, and Lecture Notes*

# LECTURE NOTES
# IN PURE AND APPLIED MATHEMATICS

1. *N. Jacobson,* Exceptional Lie Algebras
2. *L.-Å. Lindahl and F. Poulsen,* Thin Sets in Harmonic Analysis
3. *I. Satake,* Classification Theory of Semi-Simple Algebraic Groups
4. *F. Hirzebruch, W. D. Newmann, and S. S. Koh,* Differentiable Manifolds and Quadratic Forms
5. *I. Chavel,* Riemannian Symmetric Spaces of Rank One
6. *R. B. Burckel,* Characterization of C(X) Among Its Subalgebras
7. *B. R. McDonald, A. R. Magid, and K. C. Smith,* Ring Theory: Proceedings of the Oklahoma Conference
8. *Y.-T. Siu,* Techniques of Extension of Analytic Objects
9. *S. R. Caradus, W. E. Pfaffenberger, and B. Yood,* Calkin Algebras and Algebras of Operators on Banach Spaces
10. *E. O. Roxin, P.-T. Liu, and R. L. Sternberg,* Differential Games and Control Theory
11. *M. Orzech and C. Small,* The Brauer Group of Commutative Rings
12. *S. Thomeier,* Topology and Its Applications
13. *J. M. López and K. A. Ross,* Sidon Sets
14. *W. W. Comfort and S. Negrepontis,* Continuous Pseudometrics
15. *K. McKennon and J. M. Robertson,* Locally Convex Spaces
16. *M. Carmeli and S. Malin,* Representations of the Rotation and Lorentz Groups: An Introduction
17. *G. B. Seligman,* Rational Methods in Lie Algebras
18. *D. G. de Figueiredo,* Functional Analysis: Proceedings of the Brazilian Mathematical Society Symposium
19. *L. Cesari, R. Kannan, and J. D. Schuur,* Nonlinear Functional Analysis of Differential Equations: Proceedings of the Michigan State University Conference
20. *J. J. Schäffer,* Geometry of Spheres in Normed Spaces
21. *K. Yano and M. Kon,* Anti-Invariant Submanifolds
22. *W. V. Vasconcelos,* The Rings of Dimension Two
23. *R. E. Chandler,* Hausdorff Compactifications
24. *S. P. Franklin and B. V. S. Thomas,* Topology: Proceedings of the Memphis State University Conference
25. *S. K. Jain,* Ring Theory: Proceedings of the Ohio University Conference
26. *B. R. McDonald and R. A. Morris,* Ring Theory II: Proceedings of the Second Oklahoma Conference
27. *R. B. Mura and A. Rhemtulla,* Orderable Groups
28. *J. R. Graef,* Stability of Dynamical Systems: Theory and Applications
29. *H.-C. Wang,* Homogeneous Banach Algebras

*Other volumes in preparation*

# HOMOGENEOUS BANACH ALGEBRAS

Hwai-chiuan Wang

*Department of Mathematics*
*National Tsing Hua University*
*Hsinchu, Taiwan, Republic of China*

MARCEL DEKKER INC. New York and Basel

**Library of Congress Cataloging in Publication Data**

Wang, Hwai-chiuan.
Homogeneous Banach algebras.

(Lecture notes in pure and applied mathematics ; 29)
Bibliography: p.
Includes index.
1. Banach algebras. I. Title.
QA326.W35 512'.55 76-52607
ISBN 0-8247-6588-5



MARCEL DEKKER, INC.

270 Madison Avenue, New York, New York 10016

Current printing (last digit):
10 9 8 7 6 5 4 3 2 1

PRINTED IN THE UNITED STATES OF AMERICA

TO

RICHARD R. GOLDBERG

WITH WHOM I BEGAN TO LEARN HOMOGENEOUS BANACH ALGEBRAS

# PREFACE

Let G be a locally compact abelian group with character group Γ. Segal (1947) first studied the $L^1(G)$-algebras under their homogeneous structures, which are translation-invariant. However, the Segal algebras which are group algebras admitting the homogeneous structures were introduced and systematically studied by Silov (1951) and by Reiter (1968,1971). It was Katznelson (1968) who first started the investigation of the Fourier series of functions in homogeneous Banach spaces. This monograph contains a basic study of homogeneous Banach algebras, i.e., specialized homogeneous Banach spaces which are more general than Segal algebras.

The main purpose in writing this monograph is to investigate some aspects of homogeneous Banach algebras and related topics, to illustrate various methods used in several classes of group algebras, and to guide the reader toward some of the problems of current interest in harmonic analysis such as the problems of factorizations, homomorphisms, multipliers, and closed subalgebras.

The abundance and variety of examples is one of the most interesting features of the subject. Hence we complete the monograph with a table of various group algebras which will be convenient for the reader.

My first acknowledgments are due to Professor James T. Burnham, Richard R. Goldberg, and Walter Rudin for their encouragement. My thanks are also due to all the students in my classes on harmonic analysis during the academic years 1971 through 1976 for their stimulating interest. It is a pleasure to express my thanks to the Mathematics Research Center, National Science Council, Republic of China, for their financial assistance during the preparation of the manuscript. A final, special thank you is offered to Miss Judy Phone-Chie Peng, for her labor on the typescript.

Hwai-chiuan Wang

Department of Mathematics
National Tsing Hua University
Hsinchu, Taiwan
Republic of China

# CONTENTS

# CHAPTER I

# HOMOGENEOUS BANACH SPACES

In this chapter, we attempt to concern the homogeneous structures in a Banach space of functions on a locally compact abelian group $G$. As the reader will see the homogeneous structures bring about as a result of the study of the $L^1$-algebra whose fundamental properties are the chief part of section 1. In section 2, various homogeneous structures will be discussed in many directions.

## 1. THE $L^1$-ALGEBRAS

In this section, we are going to summarize the fundamental properties of $L^1(G)$ and of $\Gamma$ the character group of $G$. These properties will be used later. For a further and complete study of the $L^1$-algebras we refer to the books on harmonic analysis by Rudin (1962) and Hewitt-Ross (1970).

Unless the contrary is stated, throughout this monograph functions are complex-valued and linear space are over the complex numbers. $G$ will denote an arbitrary locally compact abelian group, and $\Gamma$ the character group of $G$. Then there exists on $G$ an essentially unique invariant Borel measure, the so-called Haar measure $dx$ on $G$. With respect to this measure there exists the space of all complex-valued integrable functions, denoted by $L^1(G)$. In $L^1(G)$, two functions coinciding almost everywhere are identified. $L^1(G)$ forms a commutative Banach algebra where the multiplication is defined by convolution, that is,

$$(f * g)(x) = \int_G f(x - y)g(y)dy$$

The norm of an element is defined by

$$\|f\|_{L^1} = \int_G |f(x)| \, dx$$

We then have

$$\|f * g\|_{L^1} \leq \|f\|_{L^1}\|g\|_{L^1} .$$

1.1 <u>THEOREM</u> Let $\Gamma$ be the character group of $G$. Then

(i) $G$ is discrete if and only if $\Gamma$ is compact.

(ii) [Pontryagin and van Kampen] The character group of $\Gamma$ is isomorphically homeomorphic to $G$.

(iii) Let $K$ and $\Delta$ be compact sets in $G$ and $\Gamma$, respectively. For $s > 0$, let

$$N(K,s) = \{\gamma \in \Gamma : |(x,\gamma) - 1| < s \text{ for } x \in K\} ,$$

$$N(\Delta,s) = \{x \in G : |(x,\gamma) - 1| < s \text{ for } \gamma \in \Delta\} .$$

Then $N(K,s)$ and $N(\Delta,s)$ are open sets in $\Gamma$ and $G$, respectively. Moreover, the family of all sets $N(K,s)$ and their translates is a base for the Gelfand topology of $\Gamma$.

1.2. <u>THEOREM</u> (i) A locally compact (Hausdorff) group is a paracompact space.

(ii) Countable compactness is equivalent to compactness in a paracompact space.

(iii) A closed subspace of a paracompact space is paracompact.

(iv) Let $G$ be a locally compact abelian group, and $E$ a non-compact closed subset of $G$. Then there exists a sequence $(x_n)$ in $E$ having no any cluster point in $G$.

(v) A locally compact abelian group is a normal space.

(vi) Let $G$ be a locally compact abelian group and $A$ any infinite subset of $G$. Then there exists a sequence $(U_n)$ of open sets whose closures are pairwise disjoint and such that $A \cap U_n \neq \phi$ for each $n$.

(vii) In a non-compact locally compact abelian group $G$, there exists a sequence $(x_n)$ in $G$ and a compact symmetric neighborhood $K$ of $0$, the identity of $G$, such that

$$(x_i + K) \cap (x_j + K) = \phi \quad \text{if} \quad i \neq j .$$

1.3. <u>THEOREM</u> $L^1(G)$ admits an identity if and only if $G$ is discrete.

1.4. <u>THEOREM</u> For $f \varepsilon L^1(G)$, define

$$\hat{f}(\gamma) = \int f(x)\, \overline{(x,\gamma)}\, dx \quad (\gamma \varepsilon \Gamma) .$$

Then (i) [Riemann-Lebesgue Lemma] $\hat{f} \in C_0(\Gamma)$ and $\|\hat{f}\|_\infty \leq \|f\|_{L^1}$, where $C_0(\Gamma)$ denotes the Banach algebra of all continuous functions on $\Gamma$ vanishing at infinity under the sup norm $\| \ \|_\infty$ with pointwise multiplication.

(ii) [Uniqueness Theorem] If $f \in L^1(G)$ with $\hat{f}(\gamma) = 0$ for all $\gamma \in \Gamma$, then $f = 0$ in $L^1(G)$.

(iii) The map $\wedge : f \to \hat{f}$ is an algebraic isomorphism of $L^1(G)$ into $C_0(\Gamma)$. That is, if $f, g \in L^1(G)$ and $a$, $b$ are complexes, then

$$(af + bg)^\wedge = a\hat{f} + b\hat{g}$$

$$(f * g)^\wedge = \hat{f}\hat{g}.$$

(iv) For $f \in L^1(G)$, $x \in G$, and $\gamma \in \Gamma$, denote the left translation operators by $L_x$, $L_\gamma$, and define $\gamma f(y) = (y,\gamma) f(y)$, $x\hat{f}(\gamma) = (x,\gamma)\hat{f}(\gamma)$, and $f^*(y) = \overline{f(-y)}$, the involution of $f$. Then

$$\widehat{L_a f} = a^{-1}\hat{f},$$

$$\widehat{\gamma f} = L_\gamma \hat{f},$$

$$\widehat{f^*} = \bar{\hat{f}}.$$

Let $L^p(G)$ be the Lebesgue space, we summarize the following results.

1.5. <u>THEOREM</u> [Plancherel] Denote the Haar measure on $G$ by $dx$. Then there exists a Haar measure $d\gamma$ on $\Gamma$, and a bijective linear map $f \to \hat{f}$ of $L^2(G)$ onto $L^2(\Gamma)$ with the following properties:

(i) $\|f\|_{L^2} = \|\hat{f}\|_{L^2}$

(ii) If $f \in L^1 \cap L^2(G)$, then $\hat{f}(\gamma) = \int f(x)\overline{(x,\gamma)}dx$ in $L^2(\Gamma)$.

1.6. THEOREM For $1 \leq p \leq 2$, there exists a linear map $f \to \hat{f}$ of $L^p(G)$ into $L^q(\Gamma)$, where $\frac{1}{q} + \frac{1}{p} = 1$, such that

(i) If $f \in L^1 \cap L^p(G)$, then $\hat{f}(\gamma) = \int f(x)\overline{(x,\gamma)}dx$ in $L^q(\Gamma)$.

(ii) [Hausdorff-Young Theorem] $\|\hat{f}\|_{L^q} \leq \|f\|_{L^p}$ $(f \in L^p(G))$.

Let $M(G)$ denote the measure algebra. For the definitions and the elementary properties of measure algebras, we refer to Hewitt-Ross [(1963, §20)], and Rudin [(1962, Ch.1 )].

1.7. THEOREM For $\mu \in M(G)$, define

$$\hat{\mu}(\gamma) = \int (-x,\gamma)\, d\mu(x) \quad (\gamma \in \Gamma) .$$

Then (i) $\hat{\mu}$ is bounded and uniformly continuous.

(ii) [Uniqueness Theorem] If $\mu \in M(G)$ with $\hat{\mu}(\gamma) = 0$ for all $\gamma \in \Gamma$, then $\mu = 0$ in $M(G)$.

(iii) The map $\wedge : \mu \to \hat{\mu}$ is an algebraic isomorphism of $M(G)$ into $C(\Gamma)$.

The linear maps in Theorems 1.4, 1.5 and 1.7 are called the Fourier transform, the Plancherel transform, and the Fourier-Stieltjes transform, respectively. However, we simply call all the linear maps in Theorems 1.4 ~ 1.7 the Fourier transforms; this based on the fact that any two different transforms $\hat{f}$ of $f$ are the same providing that $f$ belongs to the corresponding spaces. For example if $f \in L^1 \cap L^2(G)$, then the Fourier and Plancherel transforms $\hat{f}$ of $f$ are the same functions in $L^2(\Gamma)$.

1.8. <u>THEOREM</u> [Reiter (1968, pp.110-113)] Let $G$ be a locally compact abelian group with character group $\Gamma$.

(i) For any compact symmetric neighborhood $\Delta$ of $0$, the identity in $\Gamma$, there exists a generalized "triangle function" $\sigma$ in $L^1(G)$ such that

$$\sigma \geq 0 ;$$
$$\hat{\sigma}(0) = \|\sigma\|_{L^1} = 1 ;$$
$$0 \leq \hat{\sigma} \leq 1 ;$$
$$\operatorname{supp} \sigma \subset \Delta^2 .$$

(ii) For any compact symmetric neighborhoods $\Delta$ and $\Omega$ of $0$, there exists a generalized "trapezium function" $\zeta$ in $L^1(G)$ such that

$$\|\zeta\|_{L^1} \leq \left(\frac{\nu(\Delta\Omega)}{\nu(\Delta)}\right)^{\frac{1}{2}} ;$$
$$0 \leq \hat{\zeta} \leq 1 ;$$
$$\hat{\zeta}(\Omega) = 1 ;$$
$$\operatorname{supp} \hat{\zeta} \subset \Delta^2\Omega .$$

Where $\nu$ denotes the Haar measure on $\Gamma$.

(iii) If $\alpha > 1$, then there exists, for every compact $\Delta$ in $\Gamma$, a $\zeta \in L^1(G)$ such that

$$\| \zeta \|_{L^1} < \alpha ;$$
$$\hat{\zeta}(\Delta) = 1 .$$

1.9. Denote by $C_c(G)$ the set of all continuous functions on $G$ with compact support, and by $P(L^1(G))$ the set of all $f$ in $L^1(G)$ whose Fourier transforms $\hat{f}$ have compact support. Then $C_c(G)$ and $P(L^1(G))$ are dense in $L^1(G)$.

1.10. THEOREM The algebra $L^1(G)$ satisfies the following homogeneous structures:

H1. For $f \in L^1(G)$, $x \in G$, we have $L_x f \in L^1(G)$ and $\|L_x f\|_{L^1} = \|f\|_{L^1}$, where $L_x f(y) = f(y - x)$ $(y \in G)$.

H2. $x \to L_x f$ is a continuous map of $G$ into $L^1(G)$.

Afterwards the notation $f_x$ will be used as the same meaning as $L_x f$.

Let $I$ be a closed ideal of $L^1(G)$. Then $I$ determine a closed set $\Delta$ in $\Gamma$ —— its zero set $Z(I)$ where $Z(I) = \{\gamma \in \Gamma : \hat{f}(\gamma) = 0 \text{ for } f \in I\}$. Conversely, each closed set $\Delta$ in $\Gamma$ is a zero set $Z(I)$ for some closed ideal $I$ of $L^1(G)$, simply take for $I$ the set of all $f \in L^1(G)$ such that $\hat{f}(\Delta) = 0$. Can there be two distinct closed ideals $I_1$ and $I_2$ of $L^1(G)$ such that $Z(I_1) = Z(I_2)$ ? A set $\Delta$ in

$\Gamma$ such that $\Delta = Z(I)$ for a unique closed ideal $I$ of $L^1(G)$ will be called a spectral set. Is every closed set in $\Gamma$ a spectral set? If yes, we say the spectral synthesis holds for $L^1(G)$. Otherwise, we say the spectral synthesis fails for $L^1(G)$.

1.11. <u>THEOREM</u> In $L^1(G)$, the spectral synthesis holds for the case of compact $G$, but fails for the case of non-compact $G$.

A subset $\Delta$ of $\Gamma$ is a coset in $\Gamma$ if there is a subgroup $\Lambda$ of $\Gamma$ and $\gamma \in \Gamma$ such that $\Delta = \gamma + \Lambda$. In case $\Lambda$ is an open subgroup of $\Gamma$, $\Delta$ is called an open coset in $\Gamma$. By a D-coset-ring (coset-ring) of $\Gamma$, we mean the Boolean ring generated by the cosets (open cosets) in $\Gamma$.

1.12. <u>THEOREM</u> [Liu-Rooij-Wang (1973)] Let $I$ be a closed ideal in $L^1(G)$. Then $I$ admits a bounded approximate identity if and only if its zero set $Z(I)$ belongs to the D-coset-ring of $\Gamma$.

Cohen (1960) characterize the idempotent measures in the measure algebra $M(G)$ by cosets in $\Gamma$ as follows.

1.13. <u>THEOREM</u> Let $\Delta$ be a subset of $\Gamma$. Then there is a $\mu \in M(G)$ such that $\mu^2 = \mu$, $\hat{\mu}(\gamma) = 0$ for $\gamma \in \Delta$, and $\hat{\mu}(\gamma) = 1$ for $\gamma \notin \Delta$ if and only if $\Delta$ belongs to the coset-ring of $\Gamma$.

1.14. <u>THEOREM</u> (i) Suppose that $A$ and $B$ are two commutative

Banach algebras such that $B$ is semi-simple. If $\sigma$ is an algebraic isomorphism of $A$ onto $B$, then $\sigma$ is bounded.

(ii) A subalgebra of a semi-simple Banach algebra is semi-simple.

(iii) Suppose that $(B, \| \ \|_B)$ is a semi-simple commutative Banach algebra and that $A$ is a subalgebra of $B$. If $A$ is a Banach algebra under a norm $\| \ \|_A$, then there is a constant $C$ such that $\| \ \|_B \leq C \| \ \|_A$.

(iv) There is only one norm, up to the equivalence, under which a commutative Banach algebra is semi-simple.

(v) $L^1(G)$ is semi-simple.

1.15. <u>THEOREM</u> [Reiter (1968, p.20)] An ideal $I$ in $L^1(G)$ contains every function in $L^1(G)$ whose Fourier transform has compact support disjoint from the zero set of $I$. In particular, if the zero set of $I$ is empty, $I$ contains all functions in $L^1(G)$ with compactly supported Fourier transforms.

## 2. HOMOGENEOUS BANACH SPACES

In this section, our main concern is to investigate the relationship among various homogeneous structures. The principal results are how to produce a Banach space admitting more structures from a given Banach space.

2.1. <u>THEOREM</u> Let $(B(G), \| \ \|_B)$ be a Banach space of functions on $G$ admitting

H0. For $f \in B(G)$, $x \in G$, we have $f_x \in B(G)$, and $f \to f_x$ is a bounded linear operator on $B(G)$ for each $x$.

Then there exists a Banach space $(A(G), \|\ \|_A)$ as well as a linear subspace of $B(G)$ admitting

H1. For $f \in A(G)$, $x \in G$, we have $f_x \in A(G)$, and $\|f_x\|_A = \|f\|_A$.

<u>PROOF</u>. Denote by $A(G)$ the set of all functions $f$ in $B(G)$ with

$$\|f\|_A = \sup_{x \in G} \|f_x\|_B < \infty .$$

Then it is obvious that $A(G)$ is a linear subspace of $B(G)$ and $\|\ \|_A$ is a norm on $A(G)$. Moreover, $(A(G), \|\ \|_A)$ is a Banach space. Let $(f_n)$ be a Cauchy sequence in $(A(G), \|\ \|_A)$. Then $(L_x f_n)$ is Cauchy in $(B(G), \|\ \|_B)$ for each $x \in G$. In particular, $(f_n)$ is Cauchy in $(B(G), \|\ \|_B)$. Let $f \in B(G)$ with

$$f_n \to f \text{ in the } \|\ \|_B\text{-norm} .$$

Since for each $x \in G$, there exists a constant $C_x$ such that

$$\|L_x(f_n - f)\|_B \leq C_x \|f_n - f\|_B ,$$

we have

$$L_x f_n \to f_x \text{ in the } \|\ \|_B\text{-norm} ,$$

but for $\varepsilon > 0$ there exists $N > 0$ such that

(1) $$n,m \geq N \Longrightarrow \|L_x(f_n - f_m)\|_B \leq \varepsilon .$$

Passing in (1) to the limit as $m \to \infty$, we get

(2) $$\|L_x f_n - f_x\|_B \leq \varepsilon .$$

The inequality (2) is valid for all $x \varepsilon G$, so

$$\|f_n - f\|_A = \sup_{x\varepsilon G} \|L_x(f_n - f)\|_B \leq \varepsilon .$$

This is to say that $f \varepsilon A(G)$, and

$$f_n \to f \text{ in the } \|\ \|_A\text{-norm} .$$

Therefore $(A(G), \|\ \|_A)$ is a Banach space. Finally, for $f \varepsilon A(G)$, $x \varepsilon G$, we have

$$f_x \varepsilon A(G) , \text{ and } \|f_x\|_A = \|f\|_A . \qquad //$$

For many Banach spaces the H0 structure can be weaker. The exact statement follows.

2.2. <u>THEOREM</u> Let $(B(G), \|\ \|_B)$ be a Banach space of functions on $G$ satisfying the followings:

(i) Two functions are the same whenever they equal almost everywhere.

(ii) If $\|f_n - f\|_B \to 0$, then there exists a subsequence $(f_{n_k})$ of $(f_n)$ such that $f_{n_k} \to f$ a.e.

(iii) For $f \varepsilon B(G)$, $x \varepsilon G$, we have $f_x \varepsilon B(G)$.

Then for each $x \varepsilon G$ there exists a constant $C_x$ such that

$$\|f_x\|_B \leq C_x \|f\|_B \quad (f \in B(G)) .$$

<u>PROOF</u>. Fixed $x$ in $G$, define

$$|||f||| = \max\{\|f\|_B , \|f_x\|_B\}.$$

Then $|||\ |||$ is a norm on $B(G)$. Let $(f_n)$ be a Cauchy sequence in $(B(G), |||\ |||)$. Then $(f_n)$ and $(L_x f_n)$ are Cauchy in $(B(G), \|\ \|_B)$. There exists $f \in B(G)$ with

$$f_n \to f \text{ in the } \|\ \|_B\text{-norm} .$$

Let $(f_{n_k})$ be a subsequence of $(f_n)$ for which

$$f_{n_k} \to f \quad \text{a.e.}$$

There exists $g \in B(G)$ with

$$L_x f_n \to g \text{ in the } \|\ \|_B\text{-norm} ,$$

Consequently,

$$L_x f_{n_k} \to g \text{ in the } \|\ \|_B\text{-norm} .$$

Let $(f_{n_{k_\ell}})$ be a subsequence of $(f_{n_k})$ for which

$$L_x f_{n_{k_\ell}} \to g \quad \text{a.e.}$$

But

$$f_{n_{k_\ell}} \to f \quad \text{a.e.}$$

Therefore $f_x = g$ a.e. It is now clear that

$$f_n \to f \text{ in the } |||\ |||\text{-norm} .$$

Hence the normed linear space $(B(G), |||\ |||)$ is complete with

$$\|\ \|_B \leq |||\ ||| \ .$$

By the closed graph theorem, there exists a constant $C_x$ such that

$$|||\ ||| \leq C_x \|\ \|_B \ .$$

Or,

$$\|f_x\| \leq C_x \|f\|_B \qquad (f \in B(G)). \qquad //$$

We shall show below that $A(G)$ in Theorem 2.1, is equal to $B(G)$ in case $G$ is compact. Thus, for such a $B(G)$, the old norm can be changed into a new one under which $B(G)$ admits H1 .

2.3. THEOREM Let $G$ be a compact abelian group and $B(G)$ a Banach space of functions on $G$ admitting the followings:

H0. For $f \in B(G)$, $x \in G$, we have $f_x \in B(G)$, and $f \to f_x$ is a bounded linear operator on $B(G)$ for each $x$ .

H2. $x \to f_x$ is a continuous map of $G$ into $B(G)$ for each $f \in B(G)$ .

Then there exists a new norm $|||\ |||$, equivalent to the old one, under which $B(G)$ forms a Banach space admitting H0 , H1 , and H2 .

PROOF. Notations are as in Theorem 2.1, where $A(G)$ denotes the space of all $f$ in $B(G)$ for which

$$|||f||| = \sup_{x\in G} \|f_x\|_B < \infty .$$

By the hypothesis, G is compact and $x \to f_x$ is continuous for each f in B(G) ; consequently, for $f \in B(G)$ ,

$$\sup_{x\in G} \|f_x\|_B < \infty .$$

We then have A(G) = B(G) . Clearly $\| \|_B \leq ||| \; |||$ . By the closed graph theorem, $\| \|_B$ and $||| \; |||$ are equivalent. Under the norm $||| \; |||$ , B(G) admits H0 , H1 , and H2 . //

For the completeness of the study of Theorems 2.1 and 2.3, we shall give two examples of spaces on non-compact G , one shows A(G) = B(G) while the other shows $A(G) \neq B(G)$ .

2.4. EXAMPLES (i) Let R be the real group, Z the integers, and W(R) the Wiener algebra which consists of all continuous functions on R under the norm

$$\|f\| = \sum_{n\in Z} \max_{x\in[0,1]} |f(n + x)| < \infty .$$

For $f \in W(R)$ , define

$$|||f||| = \sup_{t\in R} \sum_{n\in Z} \max_{x\in[0,1]} |f(t + n + x)| .$$

Then, we have

$$\|f\| \leq |||f||| \leq 2 \|f\| .$$

It is straightforward to prove that $(W(R), \| \|)$ and $(W(R), ||| \; |||)$ are Banach algebras and is omitted. $(W(R), \| \|)$ admits H0 and H2 , but not H1. However $(W(R), ||| \; |||)$ admits H0 , H1 , and H2 .

(ii) If $G$ is non-compact and $B(G)$ any proper Beurling algebra then $A(G) \neq B(G)$ ; Corollary 5.5.

For the rest of this section, we will devote to produce a subspace admitting H1 and H2 from a space admitting H1. As a matter of fact, this subspace is "maximal" . A couple of theorems will be established, and applied to many concrete examples.

We begin with some definitions:

2.5. <u>DEFINITIONS</u> Let $(B(G), \| \ \|_B)$ be a Banach space of complex-valued measurable functions on $G$ . $B(G)$ is called a <u>semi-homogeneous Banach space on $G$</u> if the following property is satisfied.

H1. For $f \in B(G)$ , $x \in G$ , we have $L_x f \in B(G)$ and $\|L_x f\|_B = \|f\|_B$ , where $L_x f(y) = f(y - x)$ for all $y \in G$ .

If $B(G)$ satisfies the following additional property.

H2. $x \to L_x f$ is a continuous map of $G$ into $(B(G), \| \ \|_B)$ .

Then $B(G)$ is called a <u>homogeneous Banach space on $G$</u> .

Note that by H1 , H2 is equivalent to

H2'. The map $x \to L_x f$ of $G$ into $(B(G), \| \ \|_B)$ is continuous at $0$ , the identity of $G$ .

Let $B(G)$ be a semi-homogeneous Banach space. We say that $H(G)$ is a homogeneous Banach space <u>in</u> $B(G)$ if $H(G)$

is a closed linear subspace of $B(G)$ satisfying H1 and H2 under the norm $\| \ \|_B$ . $H(G)$ is the maximal homogeneous Banach space in $B(G)$ if $H(G)$ is a homogeneous Banach space in $B(G)$, and $H(G)$ contains $D(G)$ whenever $D(G)$ is any homogeneous Banach space in $B(G)$ .

If $(B(G), \| \ \|_B)$ itself is a homogeneous Banach space, then clearly $(B(G), \| \ \|_B)$ is the maximal homogeneous Banach space in $B(G)$ . Moreover,

2.6. THEOREM For every semi-homogeneous Banach space $B(G)$ , there exists the maximal homogeneous Banach space in $B(G)$ . More precisely, let $B_c(G)$ be the set of all $f$ in $B(G)$ such that the map $x \to L_x f$ of $G$ into $(B(G), \| \ \|_B)$ is continuous at $0$ . Then $B_c(G)$ is the maximal homogeneous Banach space in $B(G)$ .

PROOF. Clearly $B_c(G)$ is a linear subspace of $B(G)$ . Let $(f_n)$ be a sequence in $B_c(G)$ and $f$ in $B(G)$ with $\|f_n - f\|_B \to 0$ . For $\varepsilon > 0$ , there exists a positive integer $N$ such that $\|f_N - f\|_B < \varepsilon/3$ . Since $f_N \in B_c(G)$ , there exists a neighborhood $U$ of $0$ such that $\|L_x f_N - f_N\| < \varepsilon/3$ whenever $x \in U$ . For $x \in U$ ,

$$\begin{aligned} \|L_x f - f\|_B &= \|L_x f - L_x f_N + L_x f_N + f_N - f_N - f\|_B \\ &\le \|f - f_N\|_B + \|L_x f_N - f_N\|_B + \|f_N - f\|_B \\ &< \varepsilon . \end{aligned}$$

Hence $B_c(G)$ is closed in $B(G)$. That $B_c(G)$ satisfies H1 and H2 follows easily from the equality $\|L_{y+x}f - L_y f\|_B = \|L_x f - f\|_B$. The maximality of $B_c(G)$ is obvious. This completes that $B_c(G)$ is the maximal homogeneous Banach space in $B(G)$. //

2.7. EXAMPLES 2.7-1 ~ 2.7-3 are semi-homogeneous Banach space while 2.7-1' ~ 2.7-3' are homogeneous Banach spaces. Moreover, 2.7-i' is the maximal homogeneous Banach space in 2.7-i, for i = 1,2,3 .

2.7-1. The Lebesgue space $L^\infty(G)$.

2.7-2. The Banach space $BV(T)$ of all continuous functions $f$ on $T$ such that the total variation $V_0^{2\pi} f$ is finite, under the norm $\|f\|_{BV} = \|f\|_{L^1} + V_0^{2\pi} f$.

2.7-3. The Lipschitz space $\mathrm{Lip}_\alpha(T)$, $0 < \alpha < 1$, of all continuous functions $f$ on $T$ for which

$$\sup_{\substack{t \\ a \neq 0}} \frac{|f(t+a) - f(t)|}{|a|^\alpha} < \infty$$

under the norm

$$\|f\|_{\mathrm{Lip}_\alpha} = \|f\|_\infty + \sup_{\substack{t \\ a \neq 0}} \frac{|f(t+a) - f(t)|}{|a|^\alpha}$$

2.7-1'. The Banach space $C_b(G)$ of all bounded continuous functions on $G$ under the supremum norm.

2.7-2′. The Banach space $L^{(1)}(T)$ of all functions $f$ on $T$ such that $f$ is absolutely continuous under the norm $\|f\|_{L^{(1)}} = \|f\|_{L^1} + \|f'\|_{L^1}$ .

2.7-3′. The Banach space $lip_\alpha(T)$ , $0 < \alpha < 1$ , of all functions $f$ in $Lip_\alpha(T)$ for which

$$\lim_{a\to 0} \sup_{t} \frac{|f(t+a) - f(t)|}{|a|^\alpha} = 0$$

under the norm $\|f\|_{Lip_\alpha}$ .

It is easy to see that 2.7-1 ~ 2.7-3 are semi-homogeneous. We are going to prove $lip_\alpha(T)$ is the maximal homogeneous Banach space in $Lip_\alpha(T)$ . The other cases will be proved later.

2.8. <u>COROLLARY</u> For $0 < \alpha < 1$ , the space $lip_\alpha(T)$ is the maximal homogeneous Banach space in $Lip_\alpha(T)$ .

<u>PROOF</u>. In order to see the Corollary holds, it suffices to prove that $lip_\alpha(T) = (Lip_\alpha)_c(T)$ . Let $f \in lip_\alpha(T)$ . Then

$$\lim_{x\to 0}\|L_x f - f\|_{Lip_\alpha} = \lim_{x\to 0}\|L_x f - f\|_\infty + \lim_{x\to 0} \sup_{a\neq 0} \frac{\|L_{-a}(L_x f - f) - (L_x f - f)\|_\infty}{|a|^\alpha}$$

$$= \lim_{x\to 0} \sup_{a\neq 0} \frac{\|L_{-a}(L_x f - f) - (L_x f - f)\|_\infty}{|a|^\alpha} .$$

Since $f \in lip_\alpha(T)$ , for $\varepsilon > 0$ there exists $\delta > 0$ such that

$$|a| \le \delta \Longrightarrow \frac{\|L_{-a}f - f\|_\infty}{|a|^\alpha} < \frac{\varepsilon}{4}$$

or,

$$|a| \le \delta \Longrightarrow \frac{\|L_{-a}(L_x f-f) - (L_x f-f)\|_\infty}{|a|^\alpha} < \frac{\varepsilon}{2} \quad (x \in T) .$$

For $x \in T$ ,

$$\sup_{a\neq 0} \frac{\|L_{-a}(L_x f-f) - (L_x f-f)\|_\infty}{|a|^\alpha} \le \sup_{0<|a|<\delta} \frac{\|L_{-a}(L_x f-f) - (L_x f-f)\|_\infty}{|a|^\alpha} + \sup_{|a|>\delta} \frac{\|L_{-a}(L_x f-f) - (L_x f-f)\|_\infty}{|a|^\alpha}$$

$$< \frac{\varepsilon}{2} + \frac{2\|f_x - f\|_\infty}{\delta^\alpha} .$$

Therefore,

$$\lim_{x\to 0} \sup_{a\neq 0} \frac{\|L_{-a}(L_x f -f) - (L_x f - f)\|_\infty}{|a|^\alpha} \le \frac{\varepsilon}{2}$$

for arbitrary $\varepsilon > 0$ .

Thus,

$$\lim_{x\to 0} \|L_x f-f\|_{Lip_\alpha} = \lim_{x\to 0} \sup_{a\neq 0} \frac{\|L_{-a}(L_x f-f) - (L_x f-f)\|_\infty}{|a|^\alpha} = 0 .$$

So $f \in (Lip_\alpha)_c(T)$ .

Conversely, let $f \in (Lip_\alpha)_c(T)$ . Since $(Lip_\alpha)_c(T)$ is a homogeneous Banach space as well as a linear subspace of $L^1(T)$ , by Theorem 2.12 of Katznelson's book (1968) ,

$$\sigma_n f \to f \text{ in the } Lip_\alpha\text{-norm}$$

where $\sigma_n(f) = K_n * f$ and $(K_n)$ is the Féjer's kernel. For $\varepsilon > 0$ , choose $n > 0$ such that

$$\| \sigma_n f - f \|_{Lip_\alpha} < \frac{\varepsilon}{2} .$$

For $a, t \in T$, $a \neq 0$

$$\frac{|f(t+a)-f(t)|}{|a|^\alpha} = \frac{|f(t+a)-\sigma_n f(t+a)+\sigma_n f(t+a)-\sigma_n f(t)+\sigma_n f(t)-f(t)|}{|a|^\alpha}$$

$$\leq \frac{|(f-\sigma_n f)(t+a)-(f-\sigma_n f)(t)|}{|a|^\alpha} + \frac{|\sigma_n f(t+a)-\sigma_n f(t)|}{|a|^\alpha}$$

Thus,

$$\sup_t \frac{|f(t+a)-f(t)|}{|a|^\alpha} \leq \| f-\sigma_n f \|_{Lip_\alpha} + \sup_t \frac{|\sigma_n f(t+a)-\sigma_n f(t)|}{|a|^\alpha}$$

(3) $$< \frac{\varepsilon}{2} + \sup_t \frac{|\sigma_n f(t+a)-\sigma_n f(t)|}{|a|^\alpha}$$

But

$$\sup_t \frac{|\sigma_n f(t+a)-\sigma_n f(t)|}{|a|^\alpha} = \sup_t \frac{1}{|a|^\alpha}\left| \sum_{j=-n}^{n} (1 - \frac{|j|}{n+1})\hat{f}(j)e^{ij(t+a)} - \sum_{j=-n}^{n} (1 - \frac{|j|}{n+1})\hat{f}(j)e^{ijt}\right|$$

$$\leq \frac{\|f\|_\infty}{|a|^\alpha} \sum_{j=-n}^{n} (1 - \frac{|j|}{n+1}) |e^{ija} - 1|$$

$$\leq \frac{\|f\|_\infty}{|a|^\alpha} \sum_{j=-n}^{n} |e^{ija} - 1|$$

$$= \|f\|_\infty \sum_{j=-n}^{n} \left|\frac{e^{ija} - 1}{a^\alpha}\right| .$$

Or,

$$\lim_{a\to 0} \sup_t \frac{|\sigma_n f(t+a) - \sigma_n f(t)|}{|a|^\alpha} = \lim_{a\to 0} \|f\|_\infty \sum_{j=-n}^{n} \left|\frac{e^{ija} - 1}{a^\alpha}\right|$$

$$= \|f\|_\infty \sum_{j=-n}^{n} \lim_{a\to 0} \left|\frac{e^{ija} - 1}{|a|^\alpha}\right|$$

$$= 0 \quad \text{since} \quad \alpha < 1 .$$

So there exists $\delta > 0$ such that

$$(4) \qquad 0 < |a| < \delta \Longrightarrow \sup_t \frac{|\sigma_n f(t+a) - \sigma_n f(t)|}{|a|^\alpha} < \frac{\varepsilon}{2} .$$

Combining (3) and (4), we have

$$\lim_{a\to 0} \sup_t \frac{|f(t + a) - f(t)|}{|a|^\alpha} = 0 . \qquad //$$

Or $f \varepsilon \mathrm{lip}_\alpha(T)$ . This completes the proof. //

It seems worthwhile to state explicitly the definition of Banach $L^1(G)$-module and a factorization theorem, and then to review briefly the vector-valued integrals.

2.9. DEFINITION Let $(B, \| \|_B)$ be a Banach space and $(A, \| \|_A)$ a Banach algebra. B is called a Banach A-module if there exists a map of $A \times B$ into B , denote the image of $(f,h)$ by $f \circledast h$ , satisfying the following conditions: for $f,g \varepsilon A$ , $h,k \varepsilon B$ , and any complex number a , we have

(i) $(f + g) \circledast h = f \circledast h + g \circledast h$ , $f \circledast (h+k) = f \circledast h + f \circledast k$

(ii) $(f * g) \circledast h = f \circledast (g \circledast h)$ (* denotes the multipliaction on A )

(iii) $a(f \circledast h) = (af) \circledast h = f \circledast (ah)$

(iv) $\| f \circledast h\|_B \leq K \|f\|_A \|h\|_B$ ($K \geq 1$ a constant independent of f and h)

Hewitt (1964) and Curtis-Figà-Talamanca (1966) followed Cohen's nice method and established a decisive result:

2.10. <u>THEOREM</u> [Module Factorization Theorem] If $(A, \| \|_A)$ is a Banach algebra admitting a bounded left approximate identity [bounded by $C$] and $(B, \| \|_B)$ is a Banach $A$-module, then $A \circledast B = \{g \circledast h : g \in A , h \in B\}$ is a closed linear subspace of $(B, \| \|_B)$. More precisely, for $f$ in the closed linear span of $A \circledast B$, and $\varepsilon > 0$, there exist $g \in A$, $h \in B$ such that $f = g \circledast h$, $\|g\|_A \leq C$, $\|f - h\|_B < \varepsilon$, $h \in \overline{A \circledast f}^B$.

For a proof see Hewitt-Ross (1970, p.268).

As in the Lebesgue theory, we develop the <u>vector-valued integrals</u> as follows : Let $(B(G), \| \|_B)$ be a Banach space of measurable functions on $G$, $B^*(G)$ the dual space of $B(G)$ and $K_B(G)$ the space of all $B(G)$-valued continuous functions on $G$ with compact support. For $f \in K_B(G)$, define $\int f(x)dx$ by

$$< \int f(x)dx, h > = \int < f(x), h > dx \quad (h \in B^*(G))$$

where $x \to <f(x),h>$ is a continuous function on $G$ with compact support and $\int <f(x),h>dx$ is the Haar integral of $<f(x),h>$. Then $\int f(x)dx$ is linear on $K_B(G)$ for which $\int f(x)dx \in B(G)$, and $\| \int f(x)dx \|_B \leq \int \|f(x)\|_B dx$. Let $F_B(G)$ be the space of all $B(G)$-valued functions $f$ on $G$ for which

$$\|f\| = \overline{\int} \|f(x)\|_B dx < \infty$$

where $\overline{\int}$ is the upper integral of $\|f(x)\|_B$ in the Lebesgue sense. $\|f\|$ is a semi-norm on $F_B(G)$ under which $F_B(G)$ is complete. Let $L^1_B(G)$ be the $\| \ \|$-closure of $K_B(G)$ in $F_B(G)$. Then $L^1_B(G)$ is the space of all $B(G)$-valued measurable (with respect to $\| \ \|_B$) functions $f$ on $G$ for which $\int \|f(x)\|_B \, dx < \infty$. If we identify two functions in $L^1_B(G)$ whenever they coincide almost everywhere, then $(L^1_B(G), \| \ \|)$ forms a Banach space. Moreover, the linear operator $\int f(x)dx$ on $K_B(G)$ can be extended linearly to the space $L^1_B(G)$, also denoted by $\int f(x)dx$. For $f(x) \in L^1_B(G)$, $h \in B^*(G)$, we have

$$\left\| \int f(x)dx \right\|_B \leq \int \| f(x) \|_B \, dx ,$$

and

$$< \int f(x)dx, h> = \int <f(x), h> dx .$$

With the aid of previous arguments, we are able to prove:

2.11. <u>THEOREM</u> Any homogeneous Banach space $(B(G), \| \ \|_B)$ is a Banach $L^1(G)$-module, with suitable $\circledast$ defined as (1). Moreover $L^1(G) \circledast B(G) = B(G)$.

<u>PROOF</u>. For $f \in L^1(G)$, $h \in B(G)$. Then the map $x \to f(x)L_x h$ is a measurable function of $G$ into $(B(G), \| \ \|_B)$ with

$$\int \| f(x)L_x h \|_B \, dx < \infty .$$

This is to say $f(x)L_x h \in L^1_B(G)$. Define

$$f \circledast h = \int f(x)L_x h dx . \tag{1}$$

Then $f \circledast h \in B(G)$. For $f, g \in L_1(G)$, $h, k \in B(G)$, $a$ any complex, we have:

(i) $(f+g) \circledast h = f \circledast h + g \circledast h$ , $f \circledast (h+k) = f \circledast h + f \circledast k$

(ii) $(f * g) \circledast h = f \circledast (g \circledast h)$

(iii) $a(f \circledast h) = (af) \circledast h = f \circledast (ah)$

(iv) $\|f \circledast h\|_B \leq \|f\|_{L^1} \|h\|_B$ .

Since

$$\begin{aligned} \|f \circledast h\|_B &= \left\| \int f(x) L_x h dx \right\|_B \\ &\leq \int |f(x)| \, \|L_x h\|_B \, dx \\ &= \|f\|_{L^1} \|h\|_B \end{aligned}$$

(iv) follows. To see (ii), we prove first:

$$L_x(f \circledast h) = L_x f \circledast h = f \circledast L_x h \quad (x \in G) .$$

Let $\sigma \in B^*(G)$ , $x \in G$ . Then $\sigma \circ L_x \in B^*(G)$ , and

$$\begin{aligned} \langle L_x(f \circledast h), \sigma\rangle &= \langle f \circledast h, \sigma \circ L_x\rangle \\ &= \left\langle \int f(y) L_y h dy, \sigma \circ L_x \right\rangle \\ &= \int \langle f(y) L_y h, \sigma \circ L_x\rangle dy \\ &= \int \langle f(y) L_{y+x} h, \sigma\rangle dy \\ &= \left\langle \int f(y) L_y (L_x h) dy, \sigma \right\rangle \\ &= \langle f \circledast L_x h, \sigma\rangle . \end{aligned}$$

Similarly,

$$\begin{aligned} \langle L_x(f \circledast h), \sigma\rangle &= \int \langle f(y) L_{y+x} h, \sigma\rangle dy \\ &= \int \langle f(y-x) L_y h, \sigma\rangle dy \\ &= \langle L_x f \circledast h, \sigma\rangle . \end{aligned}$$

Thus

$$L_x(f \circledast h) = L_x f \circledast h = f \circledast L_x h .$$

Now

$$\begin{aligned} <(f * g) \circledast h,\sigma> &= \int <f * g(x)L_x h,\sigma>dx \\ &= \int f(y)g(x - y)dy \int <L_x h,\sigma>dx \\ &= \int <g(x - y)L_x h,\sigma>dx \int f(y)dy \\ &= \int f(y)<L_y g \circledast h,\sigma>dy \\ &= <\int f(y)(L_y(g \circledast h))dy,\sigma> \\ &= <f \circledast (g \circledast h),\sigma> \qquad (\sigma \in B^*(G)) \end{aligned}$$

(ii) follows. The other properties are obvious. Therefore $B(G)$ is a Banach $L^1(G)$-module. By the Module Factorization Theorem, in order to see that $L^1(G) \circledast B(G) = B(G)$ , it suffices to prove that $L^1(G) \circledast B(G)$ is dense in $(B(G),\| \ \|_B)$ . For $h \in B(G)$ . Since the map $x \to L_x h$ is continuous of $G$ into $(B(G),\| \ \|_B)$ , there exists, for $\varepsilon > 0$ , a compact neighborhood $U$ of $0$ such that

$$x \in U \Longrightarrow \|L_x h - h\|_B < \varepsilon .$$

Let $f$ be a continuous function on $G$ for which $f \geq 0$ , $\int f(x)dx = 1$ , and the support of $f$ is contained in $U$ . Then

$$\begin{aligned} \|f \circledast h - h\|_B &= \left\|\int f(x)L_x h dx - \int f(x)h dx\right\|_B \\ &\leq \int f(x) \|L_x h - h\|_B \, dx \\ &= \int_U f(x) \|L_x h - h\|_B \, dx < \varepsilon . \end{aligned}$$

This shows that $L^1(G) \circledast B(G)$ is dense in $B(G)$, and hence the theorem holds. //

2.12. **THEOREM** Let $(B(G), \| \ \|_B)$ be a semi-homogeneous Banach space. If for every $f$ in $B(G)$ the map $x \to L_x f$ is a measurable function of $G$ into $(B(G), \| \ \|_B)$, then $B(G)$ is a Banach $L^1(G)$-module, with $\circledast$ as in 2.11. Moreover, $L^1(G) \circledast B(G)$ is the maximal homogeneous Banach space in $B(G)$.

PROOF. As in the proof of Theorem 2.11, for $f \in L^1(G)$, $h \in B(G)$, define

$$f \circledast h = \int f(x) L_x h dx .$$

Then, with $\circledast$, $B(G)$ is a Banach $L^1(G)$-module. Moreover, $L_x(f \circledast h) = L_x f \circledast h = f \circledast L_x h \quad (f \in L^1(G), h \in B(G), x \in G)$. By Theorem 2.10, $L^1(G) \circledast B(G)$ is a closed linear subspace of $(B(G), \| \ \|_B)$. For $f \in L^1(G)$, $h \in B(G)$, $x \in G$, we have

$$L_x(f \circledast h) = L_x f \circledast h \in L^1(G) \circledast B(G) .$$

Thus $(L^1(G) \circledast B(G), \| \ \|_B)$ satisfies H1. H2 will follow easily from that $L^1(G)$ satisfies H2 and the followings:

$$\begin{aligned} \|L_x(f \circledast h) - f \circledast h\|_B &= \|L_x f \circledast h - f \circledast h\|_B \\ &\leq \|L_x f - f\|_{L^1} \|h\|_B \end{aligned}$$

for $f \in L^1(G)$, $h \in B(G)$, and $x \in G$. Therefore $L^1(G) \circledast B(G)$ is a homogeneous Banach space in $B(G)$. In order to see that $L^1(G) \circledast B(G)$ is the maximal homogeneous Banach space in $B(G)$, it suffices to prove that $B_c(G) \subset L^1(G) \circledast B(G)$ . Since $B_c(G)$ itself is a homogeneous Banach space, by Theorem 2.11, $L^1(G) \circledast B_c(G) = B_c(G)$ . But $L^1(G) \circledast B_c(G) \subset L^1(G) \circledast B(G)$ , so $B_c(G) \subset L^1(G) \circledast B(G)$ . //

2.13. DEFINITION A semi-homogeneous Banach space $(B(G), \| \ \|_B)$ is called convolutable (with respect to $L^1(G)$) if for $f \in L^1(G)$ , $h \in B(G)$ , then $\int f(t)h(x - t)dt$ , denoted by $f * h(x)$ , belongs to $B(G)$ for which $\|f * h\|_B \leq K\|f\|_{L^1} \|h\|_B$, where $K \geq 1$ is a constant independent of $f$ and $h$ .

2.14. REMARK Let $(B(G), \| \ \|_B)$ be a convolutable semi-homogeneous Banach space and $H(G)$ a homogeneous Banach space in $B(G)$ . If each $\sigma$ in $H^*(G)$ can be identified with a measurable function $k(x)$ on $G$ for which

$$<h,\sigma> = \int h(x)k(x)dx \quad (h \in H(G)) ,$$

then

$$f * h = f \circledast h \quad (f \in L^1(G) , h \in H(G)) .$$

For let $\sigma \in H^*(G)$ and $k(x)$ corresponds to $\sigma$ . Then

$$\begin{aligned} <f \circledast h,\sigma> &= <\int f(x)L_x h dx,\sigma> \\ &= \int <f(x)L_x h,\sigma>dx \\ &= \iint f(x)L_x h(y)k(y)dydx \\ &= \int f(x)L_x h(y)dx \int k(y)dy \\ &= \int f * h(y)k(y)dy \\ &= <f * h,\sigma> . \end{aligned}$$

This is true for all $\sigma \in H^*(G)$ . Therefore $f \circledast h = f * h$ . This proves our assertion. //

2.15. <u>THEOREM</u> Let $(B(G), \|\ \|_B)$ be a convolutable semi-homogeneous Banach space. Then $(B(G), \|\ \|_B)$ is a Banach $L^1(G)$-module, and $L^1(G)*B(G)$ is a homogeneous Banach space in $B(G)$. Moreover, suppose that $f * g = \int f(x)L_x g\,dx$ for $f \in L^1(G)$ , $g \in B_c(G)$ , then $L^1(G)*B(G)$ is the maximal homogeneous Banach space in $B(G)$ .

<u>PROOF</u>. For $f,g \in L^1(G)$ , $h,k \in B(G)$ , $x \in G$ , $a$ any complex, we have:

(i) $L_x(f * h) = L_x f * h = f * L_x h$

(ii) $(f+g)*h = f * h + g * h$ , $f * (h + k) = f * h + f * k$

(iii) $(f * g) * h = f * (g * h)$

(iv) $a(f * h) = (af) * h = f * (ah)$

(v) $\|f * h\|_B \leq K \|f\|_{L^1} \|h\|_B$ for a constant $K \geq 1$ .

All above properties follow easily from the definition of $*$ . Therefore the first conclusion holds. As in the proof of Theorem 2.12, $L^1(G)*B(G)$ is the maximal homogenoeus Banach space in $B(G)$ . //

2.16. <u>THEOREM</u> Let $(B(G), \|\ \|_B)$ be a homogeneous Banach space as well as a linear subspace of $L^1(G)$ . Then $f * h = f \circledast h$ for $f \in L^1(G)$ , $h \in B(G)$ . In particular, $B(G)$ is convolutable.

<u>PROOF</u>. Since $L^\infty(G) = L^1(G)^* \subset B^*(G)$ , for $\sigma \in L^\infty(G)$ ,

$$\begin{aligned}<f \circledast h,\sigma> &= <\int f(x)L_x h dx,\sigma> \\ &= \int <f(x)L_x h,\sigma> dx \\ &= \int dx \int f(x)L_x h(y)\sigma(y)dy \\ &= \int f(x)L_x h(y)dx \int \sigma(y)dy \\ &= \int f * h(y)\sigma(y)dy \\ &= <f * h,\sigma> \quad .\end{aligned}$$

Thus $f \circledast h = f * h$ in $L^1(G)$ , or $B(G)$ is convolutable. //

2.17. THEOREM Let $B(G)$ be a convolutable semi-homogeneous Banach space as well as a linear subspace of $L^1(G)$ . Then $L^1(G)*B(G)$ is the closure of $P(B(G))$ , where $P(B(G))$ is the set of all $h$ in $B(G)$ for which the support of the Fourier transform $\hat{h}$ of $h$ is compact. In particular, if $(B(G), \| \ \|_B)$ is a homogeneous Banach space as well as a linear subspace of $L^1(G)$ , then $B(G) = \overline{P(B(G))}^B$ .

PROOF. Let $h \in P(B(G))$ and $\Delta$ the support of $\hat{h}$ . Then $\Delta$ is compact. Let $f \in L^1(G)$ such that $\hat{f}(\Delta) = 1$ , we have $\hat{h} = \hat{f}\hat{h}$ , or $h = f * h \in L^1(G) * B(G)$ . So $P(B(G)) \subset L^1(G)*B(G)$. Next for $g \in L^1(G)$ , $h \in B(G)$ , $\varepsilon > 0$ , there exists $f \in L^1(G)$ such that the support of $\hat{f}$ is compact and $\|f-g\|_{L^1} \leq \frac{\varepsilon}{\|h\|_B+1}$. Then

$$\begin{aligned}\|f * h - g * h\|_B &\leq \|f - g\|_{L^1}\|h\|_B \\ &< \varepsilon \ .\end{aligned}$$

Therefore $L^1(G)*B(G) = \overline{P(B(G))}^B$ . In case that $B(G)$ is a homogeneous Banach space as well as a linear subspace of $L^1(G)$, then by Theorem 2.16, $B(G)$ is convolutable. Thus $B(G) = L^1(G) \circledast B(G) = \overline{P(B(G))}^B$ . //

Next we show the converse of Theorem 2.6.

2.18. <u>THEOREM</u> Let $(B(G), \| \|_B)$ be a convolutable semi-homogeneous Banach space and $(A(G), \| \|_A)$ a homogeneous Banach space <u>in</u> $B(G)$ . Suppose there is an approximate identity $E$ in $L^1(G)$ with $\|e\|_{L^1} = 1$ $(e \in E)$ and for $\varepsilon > 0$ and $g \in A(G)$ there is $e \in E$ with $\|g * e - g\|_A < \varepsilon$ . Then there is a semi-homogeneous Banach space $\tilde{A}^B$ in $B(G)$ containing $A(G)$ as a maximal homogeneous Banach space.

The notation $\tilde{A}^B$ and the proof of this theorem are presented at Theorem 6.8.

It is known that the spaces $L^\infty(G)$ , $BV(T)$ and $Lip_\alpha(T)$ are convolutable. Applying the above theorems, we have:

2.19. <u>COROLLARY</u> The maximal homogeneous Banach space in $L^\infty(G)$ is $C_b(G)$ .

<u>PROOF</u>. It is easy to see that $C_b(G)$ is a homogeneous Banach space in $L^\infty(G)$ , and $L^1(G) * L^\infty(G) \subset C_b(G)$ . By Theorem 2.15, it suffices to show that $f \circledast g = f * g$ for $f \in L^1(G)$ , $g \in (L^\infty)_c(G)$ . Since $(L^\infty)_c(G) \subset L^\infty(G)$ , $L^\infty(G)^* \subset (L^\infty)_c(G)^*$ . Let $\zeta \in L^\infty(G)^*$ , $f \in L^1(G)$ , $g \in (L^\infty)_c(G)$ ,

$$\begin{aligned}<f \circledast g,\zeta> &= \int <f(x)L_x g,\zeta>dx \\ &= \iint f(x)L_x g(y)d\zeta(y)dx \\ &= \iint f(x)L_x g(y)dxd\zeta(y) \\ &= \int f * g(y)d\zeta(y) \\ &= <f * g,\zeta> .\end{aligned}$$

Thus $f \circledast g = f * g$ . //

2.20. <u>COROLLARY</u> The maximal homogeneous Banach space in $BV(T)$ is $L^{(1)}(T)$ .

<u>PROOF.</u> Recall that, for each $f$ in $L^{(1)}(T)$ ,

$$V_0^{2\pi} f = \|f'\|_{L^1} .$$

Thus $L^{(1)}(T)$ is a closed linear subspace of $BV(T)$ . That $L^{(1)}(T)$ is a homogeneous Banach space follows form

$$(L_x f)' = L_x f' \quad (f \in L^{(1)}(T),\ x \in T) .$$

Finally we see that $L^{(1)}(T)$ is the maximal homogeneous Banach space in $BV(T)$ by proving $L^1(T) * BV(T) = L^{(1)}(T)$ . Denote $P(BV(T))$ as in Theorem 2.17. Then $P(B(G))$ is the space of all trigonometric polynomials. Since $\overline{P(BV(T))}^{L^{(1)}}(T) = L^{(1)}(T)$ (see Wang (1972)), by Theorem 2.17,

$$L^{(1)}(T) = \overline{P(BV(T))}^{L^{(1)}} = \overline{P(BV(T))}^{BV} = L^1(T) * BV(T) .$$

Thus $L^{(1)}(T)$ is the maximal homogeneous Banach space in $BV(T)$ . //

For the completeness of considering the Lipschitz spaces, we are going to investigate the maximal homogeneous Banach space in $\mathrm{Lip}_1(T)$.

It is easy to see that $\mathrm{Lip}_1(T)$ is a convolutable semi-homogeneous Banach space as well as a linear subspace of $L^1(T)$. Let $C^1(T)$ be the Banach space of all continuous functions $f$ on $T$ for which the derivative $f'$ of $f$ is also continuous under the norm

$$\|f\| = \|f\|_\infty + \|f'\|_\infty .$$

Let $f \in C^1(T)$ . Obviously, $2\|f'\|_\infty \geq \sup_{\substack{t \\ h\neq 0}} \frac{|f(t+h) - f(t)|}{|h|}$ , or $C^1(T)$ is a closed linear subspace of $\mathrm{Lip}_1(T)$ . Moreover,

2.21. <u>COROLLARY</u> $C^1(T)$ is the maximal homogeneous Banach space in $\mathrm{Lip}_1(T)$ .

<u>PROOF</u>. We also use the notation $P(\mathrm{Lip}_1(T))$ as in Theorem 2.17. Then $P(\mathrm{Lip}_1(T))$ is the set of all trigonometric polynomials, and hence $P(\mathrm{Lip}_1(T)) \subset C^1(T)$ . Thus $\overline{P(\mathrm{Lip}_1(T))}^{\mathrm{Lip}_1} \subset C^1(T)$ . But $\overline{P(\mathrm{Lip}_1(T))}^{\mathrm{Lip}_1} = C^1(T)$ (see Wang (1972)). We have, by Theroem 2.17,

$$C^1(T) = \overline{P(\mathrm{Lip}_1(T))}^{\mathrm{Lip}_1} = L^1(T) * \mathrm{Lip}_1(T) .$$

This completes the proof. //

Finally we'll see that the factorization needs not unique by Corollary 2.21 and the following theorem.

**2.22.** <u>THEOREM</u> $C^1(T) = C(T) * L^{(1)}(T)$ .

<u>PROOF.</u> By Theorem 2.11 and Corollary 2.19, $L^1(T) * C(T) = C(T)$. Let $f \in C(T)$ , $h \in L^{(1)}(T)$ , then

$$(f * h)' = f * h' \in C(T) * L^1(T) = C(T) .$$

Thus $C^1(T) \supset C(T) * L^{(1)}(T)$ . Next let $f \in C^1(T)$ . Since $f' \in C(T)$ , there exists $g \in L^1(T)$ , $h \in C(T)$ such that $f' = g * h$ . It is well-known that

$$\begin{aligned} f(x) &= f(0) + \int_0^x g * h(t)dt \quad (x \in T) \\ &= f(0) + \int g(t - y)h(y)dy \int_0^x dt \\ &= f(0) + \int_0^x g(t - y)dt \int h(y)dy . \end{aligned}$$

Let $k(x) = \int_0^x g(t)dt$ , we have

$$k'(x) = g(x) , \text{ and } (L_y k)'(x) = (L_y g)(x) .$$

Thus

$$\int_0^x g(t - y)dt = k(x - y) - k(-y)$$

or

$$f(x) = b + k * h(x)$$

where $b = f(0) - k * h(0)$ is a constant. Thus

$$\hat{k}(0)\hat{b} = \widehat{k * b}$$

or

$$\hat{k}(0)b = k * b .$$

If $\hat{k}(0) \neq 0$ , then

$$f(x) = k * \frac{b}{\hat{k}(0)}(x) + k * h(x)$$
$$= k * \left(\frac{b}{\hat{k}(0)} + h\right)(x) .$$

Thus $f = k * \left(\frac{b}{\hat{k}(0)} + h\right) \varepsilon L^{(1)}(T) * C(T)$ . In case $\hat{h}(0) \neq 0$ , we replace $k$ by $h$ in the above argument. Suppose $\hat{k}(0) = 0$ and $\hat{h}(0) = 0$ , then

$$f = b + k * h = b + (k+1) * h .$$

Using the previous argument, we have $f \varepsilon L^{(1)}(T) * C(T)$ . //

By Corollary 2.8, Theorems 2.15 and 2.16, we have

2.23. <u>COROLLARY</u> For $0 < \alpha < 1$ , $L^1(T) * Lip_\alpha(T) = lip_\alpha(T)$ .

Finally we give some remarks:

2.24. <u>REMARK</u> A Banach space $B$ is a Banach $L^1(G)$-module, by considering any one of the following two ⊛'s:

$$f \circledast g = 0 \quad \text{and} \quad f \circledast g = \left(\int f(x)dx\right)g$$

for $f \varepsilon L^1(G)$ , $g \varepsilon B$ . In the first case $L^1(G) \circledast B(G) = \{0\}$, but in the second case $L^1(G) \circledast B = B$ . Under such ⊛ , a semi-homogeneous Banach space $B(G)$ is always a Banach $L^1(G)$-module but $L^1(G) \circledast B(G)$ needs not be a maximal homogeneous Banach space in $B(G)$ .

2.25. <u>REMARK</u> We might ask if there is any non-trivial semi-homogeneous Banach space $B(G)$ in which the maximal homogeneous Banach space $B_c(G)$ is the trivial space?

Suppose that B(G) is a closed linear subspace of $L^1(G)$, then the answer is negative. For in this case, $B_c(G) = L^1(G)*B(G)$. But $L^1(G) * f = 0$ implies $f = 0$.

In general, the answer is positive.

(I) Let R, Q, and Z be the reals, the rationals, and the integers. Let $T = R/2\pi Z$, $A = Q + 2\pi Z$. Then T/A is a quotient group with uncountable elements. Denote by B(T) the linear space of the complex-valued functions f on T for which

$$f = \sum_{i=1}^{\infty} a_i \chi_{\bar{x}_i}$$

where $(\bar{x}_i)$ is a sequence in T/A and $(a_i)$ is a bounded complex sequence. Define

$$\|f\| = \sup_{t \in T} |f(t)|$$

$$= \sup_i |a_i| .$$

Then $(B(T), \|\ \|)$ forms a semi-homogeneous Banach space. The proof is a straightforward and is therefore omitted. But $B_c(T)$ is the trivial space. Suppose that

$$f = \sum_{i=1}^{\infty} a_i \chi_{\bar{x}_i} \neq 0 ,$$

and $a_N \neq 0$. Let $(b_j)$ be a sequence of irrationals in T such that

$$b_j \to 0 \quad \text{as} \quad j \to \infty .$$

$$b_j + x_i - x_N \notin A$$

for $j, i = 1,2,\cdots$. Then

$$\|L_{b_j} f - f\| = \|\sum a_i \chi_{b_j+\bar{x}_i} - \sum a_i \chi_{\bar{x}_i}\|$$

$$\geq |a_N|$$

for each $b_j$ . Therefore $f \notin B_c(T)$ , and then $B_c(T) = \{0\}$ .

Note that if we consider $B(T)$ as a linear subspace of $L^1(T)$ , then $B(T) = 0$ because each $\bar{x}$ in $T/A$ is of measure zero.

(II) (Feichtinger) Consider $G = R$ or $T$ . Let $K$ be a compact, nowhere dense subset of $G$ with positive measure, and $\chi$ the characteristic function of $K$ . Consider the set $B(G)$ of all measurable functions $g$ on $G$ such that there exists $(a_n \geq 0) \subset \ell^1$ , $(x_n) \subset G$ with $|g| \leq \sum a_n L_{x_n} \chi$ a.e. Define $\|g\|_B = \inf_{a_n} \sum_{n=1}^{\infty} a_n$ . We claim that $B(G)$ is a nontrivial semihomogeneous Banach space of measurable functions on $G$ with $B_c(G) = \{0\}$ .

(1) $B(G) \neq \{0\}$ since $\chi \in B(G)$ .

(2) $g \in B(G)$ , $x \in G \Longrightarrow L_x g \in B$ and $\|L_x g\|_B = \|g\|_B$ . Let $(a_n \geq 0) \subset \ell^1$ , $(x_n) \subset G$ satisfying $|g| \leq \sum a_n L_{x_n} \chi$ a.e. Then $|L_x g(y)| \leq \sum a_n \chi(y-x-x_n) = \sum a_n L_{x+x_n} \chi(y)$ , so $L_x g \in B(G)$ and $\|L_x g\|_B = \|g\|_B$ .

(3) Since $|g(x)| \leq \|g\|_B$ a.e., $\|g\|_B = 0$ implies $g = 0$ .

(4) $f+g \in B(G)$ and $\|f+g\|_B \leq \|f\|_B + \|g\|_B$ for $f,g \in B(G)$. Suppose that

$$|f| \leq \sum a_n L_{x_n} \chi \quad \text{a.e.}$$

$$\sum a_n \leq \|f\|_B + \frac{\varepsilon}{2}$$

$$|g| \leq \sum b_n L_{y_n} \chi \quad \text{a.e.}$$

$$\sum b_n \leq \|g\|_B + \frac{\varepsilon}{2} .$$

Define three sequences $Z_n$ , $c_n$ , $d_n$ as follows:

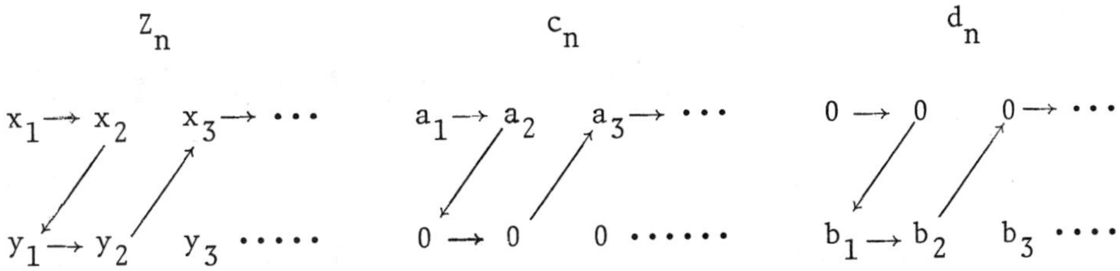


Then

$$|f| \leq \sum c_n L_{Z_n} \chi \quad \text{a.e.}$$

$$\sum c_n = \sum a_n$$

$$|g| \leq \sum d_n L_{Z_n} \chi \quad \text{a.e.}$$

$$\sum d_n = \sum b_n .$$

We have $|f+g| \leq \sum (c_n+d_n)L_{Z_n} \chi$ a.e. and $\|f+g\|_B \leq \sum(c_n+d_n)$ $\leq \|f\|_B + \|g\|_B + \varepsilon$ . Thus $f+g \in B(G)$ , and $\|f + g\|_B \leq \|f\|_B + \|g\|_B$ since $\varepsilon$ is arbitrary.

(5) $(B(G), \| \ \|_B)$ is complete.

Suppose that $(g_n)$ is a Cauchy sequence in $B(G)$ . Without loss of generality, we can assume that $\|g_n - g_{n+1}\|_B < \frac{1}{2^n}$ . Recall that $|g_n(x) - g_{n+1}(x)| \leq \|g_n - g_{n+1}\|_B$ a.e. Let

$g_n \to g$ a.e. Claim that $g_n \xrightarrow{\|\ \|_B} g$ . Take $a_{in} \geq 0$ , $x_{in} \in G$ with

$$|g_1| \leq \sum a_{1n} L_{x_{1n}} \chi \quad \text{a.e.,} \quad \sum a_{1n} < \infty$$

$$|g_1 - g_2| \leq \sum a_{2n} L_{x_{2n}} \chi \quad \text{a.e.,} \quad \sum a_{2n} \leq \frac{1}{2}$$

$$\vdots$$

$$|g_{i-1} - g_i| \leq \sum a_{in} L_{x_{in}} \chi \quad \text{a.e.,} \quad \sum a_{in} \leq \frac{1}{2^{i-1}} .$$

$$\vdots$$

Define sequences $Z_n \in G$ , $c_{in} \geq 0$ as follows:

$Z_n$ $\qquad\qquad$ $c_{in}$

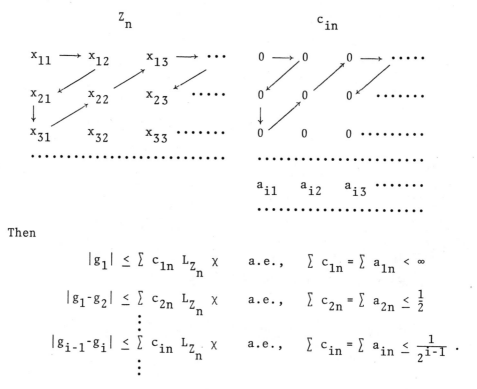

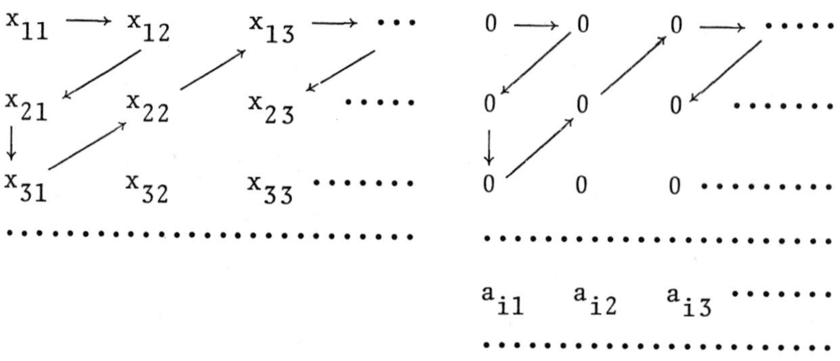


Then

$$|g_1| \leq \sum c_{1n} L_{Z_n} \chi \quad \text{a.e.,} \quad \sum c_{1n} = \sum a_{1n} < \infty$$

$$|g_1 - g_2| \leq \sum c_{2n} L_{Z_n} \chi \quad \text{a.e.,} \quad \sum c_{2n} = \sum a_{2n} \leq \frac{1}{2}$$

$$\vdots$$

$$|g_{i-1} - g_i| \leq \sum c_{in} L_{Z_n} \chi \quad \text{a.e.,} \quad \sum c_{in} = \sum a_{in} \leq \frac{1}{2^{i-1}} .$$

$$\vdots$$

We have $|g| \le |g_1| + |g_1 - g_2| + \cdots + |g_{i-1} - g_i| + \cdots$

$\le \sum_{n=1}^{\infty} (\sum_{i=1}^{\infty} c_{in}) L_{Z_n} \chi$ a.e., $\sum_{n=1}^{\infty} (\sum_{i=1}^{\infty} c_{in}) = \sum_{i=1}^{\infty} \sum_{n=1}^{\infty} c_{in}$

$\le \sum a_{1n} + \frac{1}{2} + \frac{1}{2^2} + \cdots + \frac{1}{2^{i-1}} + \cdots < \infty$ , so $g \in B(G)$ .

Furthermore,

$$\|g_m - g\|_B \le \sum_{n=1}^{\infty} \sum_{i=m+1}^{\infty} c_{in} = \sum_{i=m+1}^{\infty} \sum_{n=1}^{\infty} c_{in}$$

$$\le \frac{1}{2^m} + \frac{1}{2^{m+1}} + \cdots = \frac{1}{2^{m-1}}$$

$$\longrightarrow 0 \quad \text{as} \quad m \to \infty .$$

Hence $g_m \xrightarrow{\| \|_B} g$ .

Above arguments together with some other trivial properties reveal that $B(G)$ is a non-trivial semihomogeneous Banach space of measurable functions on $G$ .

Clearly, $B(G)$ does not contains any nonzero continuous function. $B(G) \subset L^\infty(G)$ , so $L^1(G) * B(G) \subset L^1(G) * L^\infty(G) \subset C(G)$. By Theorem 2.15, $B_c(G) = L^1(G) * B(G)$ . Therefore $B_c(G) = \{0\}$.

2.26. REMARK The theory of semi-homogeneous Banach spaces and that of homogeneous Banach spaces could be extended to a more general setting. Let $G$ be a topological group, $(B, \| \|_B)$ a Banach space, and $\mathrm{Hom}(B)$ the algebra of all bounded linear operated on $B$ . Suppose there exists a map (representation) $\sigma$ of $G$ into $\mathrm{Hom}(B)$ satisfying

H1. $\sigma(ab) = \sigma(a)\sigma(b) \quad (a,b \in G)$ ;

$\sigma(0) = \mathrm{Id}$ where $0$ , and $\mathrm{Id}$ are the identites of $G$ and $\mathrm{Hom}(B)$ , respectively;

$\|\sigma(a)\| \leq C$ , where $C \geq 1$ is a constant independent of $a$ .

Then $B$ is called a semi-homogeneous Banach space . Furthermore if $\sigma$ satisfies

H2. $x \to \sigma(x)f$ is a continuous map of $G$ into $B$ for each $f$ in $B$ .

Then $B$ is called a homogeneous Banach space. It is interesting that the theorems established in this section could be extended under these new definitions. The detail proofs in these extentions will be omitted. However, we like to point out that the proofs in these extendible theorems are the same as what have been used in the correspondent theorems.

# CHAPTER II

# SOME SPECIAL CLASSES OF BANACH ALGEBRAS

In this chapter, we intened to study some special classes of convolution algebras on a locally compact abelian group $G$, including the $C^k$-algebras, $L^p$-algebras, and many other naturally occuring group algebras. As the readers will see, these classes of algebras look very different, but share many properties.

## 3. HOMOGENEOUS BANACH ALGEBRAS

Let $G$ be a locally compact abelian group with character group $\Gamma$ . Segal (1947) first studied the $L^1$-algebras under their homogeneous structures H1 and H2. However, the Segal algebras which are group algebras admitting H1 and H2, and are generalizations of the $L^1$-algebras, was first introduced and systematically studied by the books of Reiter (1968, 1971) only a few years ago. It was Katznelson (1968) who began to study the Fourier series of functions in homogeneous Banach spaces, which are Banach spaces admitting H1 and H2, instead of the traditional study of those in the $L^1$-algebras. On the other hand, Šilov (1954) studied homogeneous rings, which are function algebras admitting H1 and H2, on a compact abelian group. This section contains a basic study of these algebras, the present author independently introduce and study under the name of homogeneous Banach algebras which are specialized homogeneous Banach spaces, but are rather more general than Segal algebras and homogeneous rings. Some, but not all, aspects of theory of Segal algebras and homogeneous rings carry over to this more general setting. It should be noticed that the maximal ideal space of a homogeneous Banach algebra is a proper subset of $\Gamma$ in general, whereas that of a Segal algebra and of a homogeneous ring must be the whole of $\Gamma$.

In addition to the richness examples of Segal algebras given in the next section, we wish to construct non-Segal homogeneous Banach algebras from Banach space and Banach algebras.

3.1. DEFINITIONS Let $G$ denote a locally compact abelian group. A homogeneous Banach algebra on $G$ is a subalgebra $B(G)$ of $L^1(G)$ such that $B(G)$ is itself a Banach algebra with respect to a norm $\| \|_B \geq \| \|_{L^1}$. It is also assumed that $(B(G), \| \|_B)$ admits H1 and H2.

A homogeneous Banach algebra is proper if it is not the whole of $L^1(G)$. A homogeneous Banach algebra is called a Segal algebra if it is dense in $L^1(G)$. A homogeneous Banach algebra $B(G)$ is a character homogeneous Banach algebra if for $f \in B(G)$, $\gamma \in \Gamma$, we have $\gamma f \in B(G)$, and $\|\gamma f\|_B = \|f\|_B$, where $\gamma f(x) = (x,\gamma)f(x)$.

The following criterion is useful.

3.2. THEOREM A homogeneous Banach space $(B(G), \| \|_B)$ is a homogeneous Banach algebra if and only if $B(G)$ is a linear subspace of $L^1(G)$ with $\| \ \|_B \geq \| \ \|_{L^1}$ .

PROOF. It suffices to consider the if part. If we denote the convolution on $L^1(G)$ by $*$ , and define,

$$f \circledast g = \int f(x)g_x dx \quad (f \in L^1(G)\ ,\ g \in B(G))\ ,$$

then, by Theorem 2.16,

$$f \circledast g = f * g \qquad (f \in L^1(G)\ ,\ g \in B(G))\ .$$

But $f \circledast g \in B(G)$ , and

$$\|f \circledast g\|_B \leq \|f\|_{L^1}\|g\|_B \quad (f \in L^1(G)\ ,\ g \in B(G))\ .$$

In particular,

$$\begin{aligned} \|f \circledast g\|_B &\leq \|f\|_{L^1}\|g\|_B \\ &\leq \|f\|_B\ \|g\|_B \quad (f,g \in B(G))\ . \end{aligned}$$

Therefore $B(G)$ is a subalgebra of $L^1(G)$ , and $(B(G), \| \ \|_B)$ forms a Banach algebra admitting H1 and H2. //

3.3. THEOREM A homogeneous Banach algebra $B(G)$ is an ideal in $L^1(G)$ . In particular, $L^1(G) * B(G) = B(G)$ .

PROOF. By Theorems 3.2, and 2.11.

Furthermore, we have

3.4. THEOREM A homogeneous Banach algebra is an ideal in the measure algebra $M(G)$ . Moreover,

$$\|\mu * f\|_B \le \|\mu\|_M \|f\|_B , \quad \text{for} \quad \mu \in M(G) , \ f \in B(G) .$$

<u>PROOF</u>. For $f \in B(G)$ , $\varepsilon > 0$ , by Theorme 2.10, there exist $g \in L^1(G)$ , $h \in B(G)$ such that

$$f = g * h$$

$$\|g\|_{L^1} \le 1$$

$$\|f - h\|_B < \varepsilon .$$

If $\mu \in M(G)$ , then $\mu * f = (\mu * g) * h \in L^1(G) * B(G) = B(G)$. This asserts that $B(G)$ is an ideal in $M(G)$ . Moreover,

$$\begin{aligned} \|\mu * f\|_B &= \|\mu * g * h\|_B \\ &\le \|\mu\|_M \|g\|_{L^1} \|h\|_B \\ &\le \|\mu\|_M \|h\|_B \\ &\le \|\mu\|_M ( \|f\|_B + \varepsilon) \end{aligned}$$

for arbitrary $\varepsilon > 0$ . So $\|\mu * f\|_B \le \|\mu\|_M \|f\|_B$ . //

Some special ideals in a homogeneous Banach algebras are ideals in $M(G)$ . Recall that an ideal $I$ in a Banach algebra $B$ is regular if there is $h \in B$ such that $fh - f \in I$ for all $f \in B$ .

3.5. <u>THEOREM</u> Any regular ideal in a homogeneous Banach algebra is an ideal in $M(G)$ .

<u>PROOF</u>. Let $B(G)$ be a homogeneous Banach algebra and $I$ a regular ideal in $B(G)$ . There exists $h \in B(G)$ with

$f * h - f \in I$ for all $f \in B(G)$. Suppose $\mu \in M(G)$, $f \in I$, we have

$$\mu * f * h - \mu * f \in I .$$

But $\mu * f * h \in I$, so $\mu * f \in I$. //

3.6. <u>THEOREM</u> Let $B(G)$ be a homogeneous Banach algebra admitting a weak approximate identity. Then a closed ideal in $B(G)$ is an ideal in $M(G)$.

<u>PROOF</u>. For $\mu \in M(G)$, $f \in I$, $\varepsilon > 0$, there exists $h \in B(G)$ with

$$\|h * f - f\|_B < \frac{\varepsilon}{\|\mu\|_M + 1} .$$

Therefore

$$\|\mu * h * f - \mu * f\|_B < \varepsilon .$$

Since $\mu * h * f \in I$ and $\varepsilon > 0$ is arbitrary, $\mu * f \in I$. //

The following important result is valid.

3.7. <u>THEOREM</u> Let $B(G)$ denote a homogeneous Banach algebra. Then (i) If $P(B(G))$ denotes the space of all functions $f$ in $B(G)$ whose Fourier transforms are of compact support, then $P(B(G))$ is dense in $B(G)$.

(ii) If $E$ is a dense ideal in $L^1(G)$, then $B(G) \cap E$ is dense in $B(G)$.

(iii) If $(k_\lambda)$ is an approximate identity in $L^1(G)$, then

$$\|k_\lambda * f - f\|_B \to 0 \quad \text{for every } f \in B(G) .$$

(iv) If $G$ is compact, then $B(G)$ admits an approximate identity.

PROOF. (i) By Theorem 2.17.

(ii) For $f \in B(G)$, $\varepsilon > 0$, there exists, by Theorem 2.10, $g \in L^1(G)$, $h \in B(G)$ with

$$f = g * h$$

$$\|f - h\|_B < 1 .$$

Since $E$ is dense in $L^1(G)$, there exists $k \in E$ such that

$$\|k - g\|_{L^1} < \frac{\varepsilon}{\|h\|_B + 1} .$$

Thus

$$\begin{aligned} \|k * h - f\|_B &= \|k * h - g * h\|_B \\ &\leq \|k - g\|_{L^1}\|h\|_B \\ &< \varepsilon . \end{aligned}$$

But $k * h \in B(G) \cap E$. This proves our assertion.

(iii) For $f \in B(G)$, there are $g \in L^1(G)$, $h \in B(G)$ with $f = g * h$. Since $(k_\lambda)$ is an approximate identity in $L^1(G)$,

$$\begin{aligned} \|k_\lambda * f - f\|_B &= \|k_\lambda * g * h - g * h\|_B \\ &\leq \|k_\lambda * g - g\|_{L^1}\|h\|_B \\ &\to 0 \text{ as } \lambda \to \infty . \end{aligned}$$

(iv) Let $(k_\lambda)$ be an approximate identity in $L^1(G)$

with compactly supported Fourier transforms. For each $\lambda$, if $k_\lambda = \sum_{i=1}^{n} c_i\gamma_i$, then set $k'_\lambda = \sum_{i=1}^{n}{}' c_i\gamma_i$, where $\sum_{i=1}^{n}{}'$ is the sum of all $c_i\gamma_i$, $i = 1,2,\cdots, n$ in which $\gamma_i$ are outside the zero set $Z(B(G))$ of $B(G)$. By Theorem 1.18, $k'_\lambda \in B(G)$. Moreover $k_\lambda * f = k'_\lambda * f$ for each $\lambda$ and each $f \in B(G)$. Hence, by (iii),

$$\|k'_\lambda * f - f\|_B = \|k_\lambda * f - f\|_B \to 0 \quad \text{as} \quad \lambda \to \infty.$$

This proves the assertion. //

We prove the following result which is a generalization of the Riemann-Lebesgue lemma.

3.8. <u>THEOREM</u> Let $B(T)$ be a homogeneous Banach algebra on $T$. For every $f \in B(T)$, and every $e^{int}$, we have $\hat{f}(n)e^{int} \in B(T)$ and $\lim_{n\to\infty} \|\hat{f}(n)e^{int}\|_B = 0$. In particular, if $B(T) = L^1(T)$, it is the Riemann-Lebesgue lemma.

<u>PROOF</u>. For every $f \in B(T)$, and every $e^{int}$, we have

$$\hat{f}(n)e^{int} = f * e^{int} \in B(T) .$$

Since $e^{int} \in L^1(T)$ and $B(T)$ is an ideal in $L^1(T)$. Suppose $f \in B(T)$, and $\varepsilon > 0$, by Theorem 3.7, there exists a trigonometric polynomial $g$ in $P(B(T))$ of degree $N$, say, such that

$$\|g - f\|_B < \varepsilon .$$

If $n \geq N + 1$ , then $g * e^{int} = 0$ . We conclude that

$$\begin{aligned} \|\hat{f}(n)e^{int}\|_B &= \|f * e^{int}\|_B \\ &= \|f * e^{int} - g * e^{int}\|_B \\ &\leq \|f - g\|_B \|e^{int}\|_{L^1} \\ &< \varepsilon . \end{aligned}$$

This asserts that $\lim_{n\to\infty} \|\hat{f}(e)^{int}\|_B = 0$ . //

One of the applications of the Module Factorization Theorem follows:

3.9. <u>THEOREM</u> Let $B(G)$ be a homogeneous Banach algebra admitting a bounded approximate identity. Then $B(G)$ is closed in $L^1(G)$ .

<u>PROOF</u>. By the hypothesis, $B(G)$ admits a bounded approximate identity. As a Banach $B(G)$-module, $L^1(G)$ contains $B(G) * L^1(G)$ ( $= B(G)$) as a closed subspace. Therefore $B(G)$ is closed in $L^1(G)$ , or both $\| \ \|_B$ and $\| \ \|_{L^1}$ are equivalent. //

Suppose $(B(G), \| \ \|_B)$ is a homogeneous Banach algebra. For any non-empty set $\Delta$ in $\Gamma$ , let $B_\Delta(G)$ be the linear space of all functions $f$ in $B(G)$ for which $\hat{f}(\Delta^c) = 0$ , where $\Delta^c$ is the complement of $\Delta$ . Under the norm $\| \ \|_B$ , $B_\Delta(G)$ forms a non-Segal homogeneous Banach algebra. Such a kind of homogeneous Banach algebra exhausts the homogeneous

Banach algebra admitting a bounded approximate identity in case $G$ is compact.

3.10. THEOREM Let $G$ be a compact abelian group. If $B(G)$ is a homogeneous Banach algebra admitting a bounded approximate identity, then the zero set $\Delta$ of $B(G)$ belongs to the coset-ring of $\Gamma$, and $B(G) = L^1_{\Delta^c}(G)$. Conversely, for any $\Delta$ in the coset-ring of $\Gamma$, $L^1_{\Delta^c}(G)$ is a homogeneous Banach algebra admitting a bounded approximate identity.

PROOF. First assume $B(G)$ is a homogeneous Banach algebra admitting a bounded approximate identity, whose zero set is $\Delta$. Then by Theorem 3.9, $B(G)$ is a closed ideal in $L^1(G)$. Both $B(G)$ and $L^1_{\Delta^c}(G)$ are closed ideals in $L^1(G)$ with the same zero set. Since the spectral synthesis holds for $L^1(G)$ in case $G$ is compact, $B(G) = L^1_{\Delta^c}(G)$. The other part follows easily from Theorem 1.12.

3.11. REMARK By Theorem 1.15, homogeneous Banach algebra $B(G)$ contains all functions $f$ in $L^1(G)$ whose Fourier transforms $\hat{f}$ are of compact support disjoint from the zero set $Z(B(G))$ of $B(G)$.

For any two homogeneous Banach algebras $B_1(G)$ and $B_2(G)$ such that $B_1(G) \subset B_2(G)$, then, by Theorem 1.14,

$$\| \quad \|_{B_2} \leq C\| \quad \|_{B_1}$$

for some constant $C$. Denote by $Z(B_1)$ and $Z(B_2)$ the zero sets of $B_1(G)$ and $B_2(G)$, respectively. Obviously,

$$Z(B_2) \subset Z(B_1) .$$

Let $\Delta$ be any compact set in $\Gamma$ disjoint from $Z(B_1)$. For any $f$ in $B_2(G)$ with $\text{Supp}\,\hat{f} \subset \Delta$, we have

$$f \in B_1(G) .$$

Moreover,

3.12. <u>THEOREM</u> Let $B_1(G)$ and $B_2(G)$ be two homogeneous Banach algebras with $B_1(G) \subset B_2(G)$. Let $\Delta$ be a compact set in $\Gamma$ disjoint from $Z(B_1(G))$, then there exists a constant $C$ such that

$$\|f\|_{B_1} \leq C\,\|f\|_{B_2}$$

for all $f$ in $B_1(G)$ with $\text{Supp}\,\hat{f} \subset \Delta$.

<u>PROOF</u>. Since $Z(B_1)$ is closed and $\Delta \cap Z(B_1) = \phi$, by Theorem 1.11 (ii) there exists a function $g \in L^1(G)$ such that the support of $\hat{g}$ is compact,

$$\hat{g}(\Delta) = 1$$

$$\hat{g}(Z(B_1)) = 0 .$$

Such a function $g$ belongs to $B_1(G)$, and

$$f = g * f$$

for every $f$ in $B_1(G)$ with $\text{Supp}\,\hat{f} \subset \Delta$. Hence,

$$\begin{aligned}\|f\|_{B_1} &= \|g * f\|_{B_1} \\ &\leq \|g\|_{B_1} \|f\|_{L^1} \\ &\leq \|g\|_{B_1} \|f\|_{B_2} .\end{aligned}$$

The theorem follows by letting $C = \|g\|_{B_1}$ . //

In the rest of this section, we devote to construct non-Segal homogeneous Banach algebras:

## 3.13. EXAMPLES

3.13-1. For any two homogeneous Banach algebras $B_1(G)$ and $B_2(G)$ , the algebra $B_1 \cap B_2(G)$ under the sum norm is a homogeneous Banach algebra.

3.13-2. Let $B(G)$ be a semi-homogenoeus Banach space such that $B_c(G)$ is a linear subspace of $L^1(G)$ with $\| \ \|_B \geq \| \ \|_{L^1}$ , then such a $B_c(G)$ is a homogeneous Banach algebra.

3.13-3. Let $(B(G), \| \ \|_B)$ be a Banach space of measurable functions on $G$ admitting H0 . Denote

$$A(G) = \{f \in B(G): \|f\|_A = \sup_{x \in G} \|f_x\|_B < \infty\}.$$

Then $(A(G), \| \ \|_A)$ is a semi-homogeneous Banach space. If $A_c(G)$ is a linear subspace of $L^1(G)$ with $\| \ \|_A \geq \| \ \|_{L^1}$ , then it is a homogeneous Banach algebra.

3.13-4. The homogeneous Banach algebras $B_\Delta(G)$ (see the paragraph after Theorem 3.9).

3.13-5. For, $\mu \in M(G)$ , the algebra $\mu * L^1(G)$ of all functions $\mu * f$ , where $f \in L^1(G)$ , under the norm $\|\mu * f\| = \|f\|_{L^1}$ .

Then $\mu * L^1(G)$ is a homogeneous Banach algebra. If $L^1(G)$ is replaced by any homogeneous Banach algebra $B(G)$ , then $\mu * B(G)$ still be a homogeneous Banach algebra.

3.13-6. [Johnson (1973)] For $f \in L^1(R)$ , the algebra $S_f$ of all functions $g$ in $L^1(R)$ with $f * g \in C_0(R)$ under the norm $\|g\| = \|g\|_{L^1} + \|f * g\|_\infty$ , then $S_f$ is a homogeneous Banach algebra. Here we could replace $L^1(R)$ by any homogeneous Banach algebra.

3.13-7. Let $G$ be a locally compact abelian group with character group $\Gamma$ , let $\mu$ be a positive unbounded regular measure on $\Gamma$ , and let $p$ be with $1 \leq p < \infty$ . Recall that a regular measure $\mu$ on a locally compact space $X$ is a measure on a $\sigma$-algebra of subsets of $X$ containing the $\sigma$-algebra of Borel sets of $X$ such that

(i) $\mu(K) < \infty$ for every compact set $K$ in $X$ :

(ii) $\mu(A) = \inf \{ \mu(U) : U$ is open in $X , A \subset U \}$ for every measurable set $A$ in $X$ .

(iii) $\mu(U) = \sup \{ \mu(K) : K$ is compact in $X, K \subset U\}$ for open sets $U$ in $X$ .

For any homogeneous Banach algebra $(B(G), \| \ \|_B)$ , let $B^{(p,\mu)}(G)$ be the linear space of all functions $f$ in $B(G)$ with $\hat{f} \in L^p(\mu)$ . Then under the norm

$$\|f\| = \|f\|_B + \|\hat{f}\|_{L^p(\mu)}$$

$B^{(p,\mu)}(G)$ forms a normed linear space as well as a linear subspace of $L^1(G)$ with $\| \ \| \geq \| \ \|_{L^1}$ . The verfications are straightforward and are therefore omitted. Moreover,

3.13-7-1. THEOREM $(B^{(p,\mu)}(G), \|\ \|)$ is a homogeneous Banach algebra.

PROOF. I. $B^{(p,\mu)}(G)$ is a Banach space.

Let $(f_n)$ be a Cauchy sequence in $B^{(p,\mu)}(G)$, then $(f_n)$ and $(\hat{f}_n)$ are Cauchy sequences in $B(G)$ and $L^p(\mu)$, respectively. There exists $f \in B(G)$ and $g \in L^p(\mu)$ such that $f_n \to f$ in $B$ and $\hat{f}_n \to g$ in $L^p(\mu)$. We have that $\hat{f}_n$ uniformly converges to $\hat{f}$. Furthermore, there exists a subsequence $(\hat{f}_{n_k})$ of $(\hat{f}_n)$, which converges almost everywhere to $g$. Thus $\hat{f} = g$ in $L^p(\mu)$, so $f \in B^{(p,\mu)}(G)$, and $\|f_n - f\| = \|f_n - f\|_B + \|\hat{f}_n - \hat{f}\|_{L^p(\mu)} \to 0$ as $n \to \infty$. Therefore $B^{(p,\mu)}(G)$ is a Banach space.

II. $B^{(p,\mu)}(G)$ satisfies H1.

For $f \in B^{(p,\mu)}(G)$, $a \in G$,

$$\begin{aligned} \|L_a f\| &= \|L_a f\|_B + \|\widehat{L_a f}\|_{L^p(\mu)} \\ &= \|f\|_B + \|\overline{(a,\gamma)}\hat{f}\|_{L^p(\mu)} \\ &= \|f\|_B + \|\hat{f}\|_{L^p(\mu)} \\ &= \|f\| . \end{aligned}$$

III. $B^{(p,\mu)}(G)$ satisfies H2'.

That is to show that $x \to L_x f$ is continuous at the identity $0$. For $f \in B^{(p,\mu)}(G)$, $\varepsilon > 0$, there exists a neighborhood $U_B$ of $0$ such that $x \in U_B$ implies $\|L_x f - f\|_B < \varepsilon/2$.

Since $\hat{f} \in L^p(\mu)$ there exists a compact set $\Delta$ in $\Gamma$ such that $(\int_{\Delta^c} |\hat{f}|^p d\mu)^{1/p} < \varepsilon/8$ . Let $M = \|f\|_{L^p(\mu)} + 1$ , $U_p = \{x \in G : |(x,\gamma) - 1| < \varepsilon/4M$ for all $\gamma \in \Delta\}$ . Then $U_p$ is a neighborhood of $0$ . For $s \in U_p$ , we have

$$\begin{aligned}
\|\widehat{L_x f - f}\|^p_{L^p(\mu)} &= \|\overline{(x,\gamma)}\hat{f} - \hat{f}\|^p_{L^p(\mu)} \\
&= \int_\Gamma |\overline{(x,\gamma)} - 1|^p |\hat{f}|^p d\mu \\
&= \int_\Delta |(x,\gamma) - 1|^p |\hat{f}|^p d\mu \\
&\quad + \int_{\Delta^c} |(x,\gamma) - 1|^p |\hat{f}|^p d\mu \\
&\le (\tfrac{\varepsilon}{4M})^p \|\hat{f}\|^p_{L^p(\mu)} + 2^p(\varepsilon/8)^p \\
&< (\varepsilon/4)^p + (\varepsilon/4)^p \\
&= 2(\varepsilon/4)^p \\
&\le (\varepsilon/2)^p .
\end{aligned}$$

Thus $\|\widehat{L_x f - f}\|_{L^p(\mu)} \le \varepsilon/2$ . Let $U = U_B \cap U_p$ . Then $U$ is a neighborhood of $0$ such that, for $x \in U$ , we have

$$\begin{aligned}
\|L_x f - f\| &= \|L_x f - f\|_B + \|\widehat{L_x f - f}\|_{L^p(\mu)} \\
&< \varepsilon/2 + \varepsilon/2 \\
&= \varepsilon .
\end{aligned}$$

Combining I , II , and III , $B^{(p,\mu)}(G)$ forms a homogeneous Banach algebra. //

$B^{(p,\mu)}(G)$ will be simply denoted by $B^p(G)$ if $\mu$ is the Haar measure on $\Gamma$, and by $(L^1)^p(G)$ or $A^p(G)$ in case $B(G) = L^1(G)$ and $\mu$ is the Haar measure. The theory of the $A^p$-algebras was first studied by Iwasawa (1944) and Larsen-Liu-Wang (1964) and then subsequently studied by many other authors.

3.13-7-2. <u>THEOREM</u> Let $B(G)$ be a homogeneous Banach algebra, $\mu$ a positive unbounded regular measure, and $1 \leq p < \infty$.

(i) Suppose that $B(G)$ is character, then $B^p(G)$ is character.

(ii) Suppose that $B(G)$ admits an <u>approximate identity</u> $(k_\lambda)$ with compactly supported Fourier transforms, then $B^{(p,\mu)}(G)$ also admits $(k_\lambda)$ as an approximate identity.

(iii) Suppose that $B^{(p,\mu)}(G)$ is dense in $B(G)$, and admits a <u>bounded approximate identity</u> $(k_\lambda)$ with compactly supported Fourier transforms, then $B(G)$ admits $(k_\lambda)$ as a bounded approximate identity.

<u>PROOF.</u> (i) and (ii) are obvious. For (iii), let $(k_\lambda)$ be an approximate identity in $B^{(p,\mu)}(G)$ bounded by $M$, whose Fourier transform $\hat{k}_\lambda$ are of compact support. For $f \in B(G)$ and $\varepsilon > 0$ there exists $g \in B^{(p,\mu)}(G)$ such that $\|g - f\|_B < \varepsilon/2(M + 1)$. For such $g$, there exists $\lambda_0$ such that $\|k_\lambda * g - g\|_B < \varepsilon/2$ whenever $\lambda \geq \lambda_0$.

For $\lambda \geq \lambda_0$, we have

$$\begin{aligned}\|k_\lambda * f - f\|_B &\le \|k_\lambda * f - k_\lambda * g + k_\lambda * g - g + g - f\|_B \\ &\le \|k_\lambda * (f-g)\|_B + \|k_\lambda * g - g\|_B + \|g - f\|_B \\ &< (M+1)\,\|g - f\|_B + \|k_\lambda * g - g\|_B \\ &< \varepsilon .\end{aligned}$$

Thus $(k_\lambda)$ is a bounded approximate identity in $B(G)$ . //

As the readers will see, it is very useful to consider the homogeneous Banach algebras $B^{(p,\mu)}(G)$ in case that $\mu$ is discrete. Let $(a_n)$ be a sequence in $\Gamma$ , and let

$$\mu = \sum_{n=1}^{\infty} c_n \delta_{a_n}$$

where $\delta_{a_n}$ is the Dirac measure concertrated at $a_n$ and $c_n$ is a complex number for each $n$ . It is unnecessary for such a $\mu$ to be regular. For example, if $(a_n)$ has a cluster point and $c_n = 1$ for each $n$ , then $\mu$ is not regular. However, $\mu$ is regular whenever $(a_n)$ is an infinite sequence which has no any cluster points. Such a sequence $(a_n)$ could be picked up from any non-compact closed set in $\Gamma$ [Theorem 1.2(iv)].

3.13-8. Let $B_1(G)$ and $B_2(G)$ be any two homogeneous Banach algebras, and $\xi$ a continuous function on $\Gamma$ . $B_1^{(\xi,B_2)}(G)$ denotes the space of all functions $f$ in $B_1(G)$ such that $\hat{f}\xi \in \widehat{B_2(G)}$ . For $f \in B_1^{(\xi,B_2)}(G)$ , define $\sigma(f) \in B_2(G)$ with $\widehat{\sigma(f)} = \hat{f}\xi$ , and then define

$$\|f\| = \|f\|_{B_1} + \|\sigma(f)\|_{B_2} .$$

By the linearity of $\sigma$ , $\| \ \|$ is a norm on $B_1^{(\xi,B_2)}(G)$ .

Moreover,

3.13-8-1. <u>THEOREM</u> $(B_1^{(\xi,B_2)}(G), \| \ \|)$ is a homogeneous Banach algebra.

<u>PROOF.</u> I. $(B_1^{(\xi,B_2)}(G), \| \ \|)$ is a Banach space.

Let $(f_n)$ be a Cauchy sequence in $(B_1^{(\xi,B_2)}(G), \| \ \|)$ , then $(f_n)$ and $(\sigma(f_n))$ are Cauchy in $B_1(G)$ and $B_2(G)$ , respectively. Let $f \in B_1(G)$ , $g \in B_2(G)$ such that

$$\|f_n - f\|_{B_1} \to 0$$

and

$$\|\sigma(f_n) - g\|_{B_2} \to 0 .$$

Thus $\hat{f}_n \to \hat{f}$ and $\widehat{\sigma(f_n)} \to \hat{g}$ . This asserts that $\hat{g} = \hat{f}\xi$ , or $g = \sigma(f)$ . It follows that $f_n$ converges to $f$ in $(B_1^{(\xi,B_2)}(G), \| \ \|)$ .

II. $(B_1^{(\xi,B_2)}(G), \| \ \|)$ satisfies H1 and H2 .

This assertion follows from the identity

$$\sigma(f_x) = (\sigma(f))_x$$

for every $f \in B_1^{(\xi,B_2)}(G)$, $x \in G$ . //

3.13-8-2. <u>REMARKS</u> For any three homogeneous Banach algebras $B(G)$ , $B_1(G)$ , $B_2(G)$ with $B_1(G) \subset B_2(G)$ , if $\xi$ is a continuous function on $\Gamma$ , then

(i) $B_1^{(\xi,B)}(G) \subset B_2^{(\xi,B)}(G)$ .

(ii) $B^{(\xi,B_1)}(G) \subset B^{(\xi,B_2)}(G)$ .

(iii) $B = B^{(\hat{\mu},B)}(G)$ where $\hat{\mu}$ is a Fourier-Stieltjes transform.

(iv) $B_1^{(1,B_2)}(G) = B_1 \cap B_2(G)$ .

In the final of this section, we are attempting to give a somewhat natural example of homogeneous Banach algebra which will be illustrated by the following results:

3.13-9-1. THEOREM Let $B(G)$ be a homogeneous Banach algebra. If $A$ is a closed invariant subspace of $B(G)$ , then, under the $B(G)$-norm, $A$ forms a homogeneous Banach algebra. In particular, $A$ is a closed ideal of $B(G)$ .

PROOF. By Theorems 3.2 and 3.3. //

Note that such a closed invariant subspace $A$ above is also an ideal of $L^1(G)$ by Theorem 3.3, but, nevertheless, every closed ideal of a homogeneous Banach algebra $B(G)$ is unnecessarily a closed invariant subspace of $B(G)$ (compare with Theorem 4.13 ). However, partial converse of the preceding theorem is acquired:

3.13-9-2. THEOREM Suppose that $B(G)$ is a homogeneous Banach algebra admitting a weak approximate identity, then any closed ideal of $B(G)$ is a closed invariant subspace of $B(G)$ .

PROOF. Let $I$ be a closed ideal of $B(G)$ and $f \in I$ . Since $B(G)$ admits a weak approximate identity, there exists, for $\varepsilon > 0$ , $g \in B(G)$ such that

$$\|g * f - f\|_B < \varepsilon .$$

Consequently,

$$||L_x(g * f) - L_x f||_B < \varepsilon \qquad (x \in G)$$

But $L_x(g * f) = L_x g * f$ , it truns out

$$||L_x g * f - L_x f||_B < \varepsilon$$

$L_x f \in I$ follows from $L_x g * f \in I$ , $\varepsilon$ is arbitrary and $I$ is closed. This proves our assertion. //

3.13-9-3. <u>COROLLARY</u> Suppose that $B(G)$ is a homogeneous Banach algebra. If $G$ is compact, then every closed ideal of $B(G)$ is a closed invariant subspace of $B(G)$ .

<u>PROOF.</u> By Theorems 3.7 (iv) , and 3.13-9-2. //

3.13-9-4. <u>THEOREM</u> Suppose that $B(G)$ is a homogeneous Banach algebra. If $I$ is a regular ideal of $B(G)$ , then $I$ is an invariant subspace of $B(G)$ .

<u>PROOF.</u> Suppose that $I$ is a regular ideal of $B(G)$ . Pick $h \in B(G)$ such that $f * h - f \in I$ . In particular, $g \in I$ , $x \in G$ , $L_x g * h - L_x g \in I$ . But $L_x g * h = g * L_x h \in I$ , so $L_x g \in I$ . Therefore $I$ is invariant.

## 4. SEGAL ALGEBRAS AND FP-ALGEBRAS

In this section, we wish to derive the fundamental properties from the results of the last section, to give many examples of Segal algebras, and to introduce and study the FP-algebras.

The following basic theorems for Segal algebras are easy consequences of the last section or referable to Reiter's books (1968, 1971).

4.1. THEOREM

(i) Let $S_1(G)$ and $S_2(G)$ be two Segal algebras. Then $S_1 \cap S_2(G)$ is a Segal algebras under the sum norm

$$||f|| = ||f||_{S_1} + ||f||_{S_2} .$$

(ii) Let $S_1(G)$ and $S_2(G)$ be two Segal algebras with $S_1(G) \subset S_2(G)$, and $\Delta$ a compact set in $\Gamma$. Then, there exists a constant $C$, such that

$$||f||_{S_1} \leq C\, ||f||_{S_2}$$

for every $f$ in $S_1(G)$ such that the support of Fourier transform $\hat{f}$ of $f$ is contained in $\Delta$.

4.2. THEOREM

(i) For any Segal algebra $S(G)$, $P(L^1(G))$ is contained in $S(G)$. In particular, $P(L^1(G)) = P(S(G))$.

(ii) For any Segal algebras $S(G)$, the space $P(L^1(G))$ is dense in $S(G)$.

(iii) If $G$ is discrete, then any Segal algebras $S(G)$ is the whole of $L^1(G)$.

(iv) [Reiter (1971,p.26)] The intersection of all Segal algebras $S(G)$ is precisely $P(L^1(G))$.

4.3. <u>THEROEM</u> [Reiter (1968,p.129)] Suppose that $S(G)$ is a Segal algebra. If $I$ is a closed ideal in $S(G)$, then the closure $\bar{I}^{L^1}$ of $I$ in $L^1(G)$ is a closed ideal in $L^1(G)$ such that $I = \bar{I}^{L^1} \cap S(G)$. Moreover, if $J$ is any closed ideal in $L^1(G)$ such that $I = J \cap S(G)$, then $J = \bar{I}^{L^1}$.

4.4. <u>THEOREM</u>

(i) A proper Segal algebra does not admit any bounded approximate identity.

(ii) Let $S(G)$ be a Segal algeebra, then

$$L^1(G) * S(G) = S(G)$$

(iii) A Segal algebra admits an approximate identity.

<u>PROOF.</u> (i) and (ii) follow from Theorem 3.9 and Theroem 3.3, respectively. (iii) follows from Theorems 3.7 (iii) and 4.2(i).

The Segal algebras are closely related to the FP-algebras:

4.5. <u>DEFINITIONS</u> Let $G$ be a locally compact abelian gorup with character group $\Gamma$, and $\mu$ a positive unbounded regular measure on $\Gamma$. A Banach algebra $(B(G), \|\ \|_B)$ in $L^1(G)$ is an <u>$F^\mu$-algebra</u> if $\widehat{B(G)} \subset L^p(\mu)$ for some $p$, $0 < p < \infty$. A Banach algebra $(B(G), \|\ \|_B)$ in $L^1(G)$ is a <u>$P^\mu$-algebra</u> if there exist two sequences $(\Delta_n)$, and $(\theta_n)$ in $\Gamma$, a sequence $(f_n)$ in $B(G)$, and a sequence $(C_n \geq 1)$ satisfying

P-1. $\Delta_i \cap \Delta_j = \phi$ if $i \neq j$, $\theta_n \subset \text{Int}(\Delta_n)$, $\mu(\theta_n) = \alpha > 0$, $\mu(\Delta_n) = \beta < \infty$ for $n = 1,2,\cdots$, (Here, Int denotes interior).

P-2. $0 \le \hat{f}_n \le 1$ , Supp $\hat{f}_n \subset \Delta_n$ , $\hat{f}_n(\theta_n) = 1$ for $n = 1,2,\cdots$ .

P-3. $\|f_n\|_B \le C_n$ , $\sum_{n=1}^{\infty} \frac{1}{C_n^a} < \infty$ , $\sum_{n=1}^{\infty} \frac{1}{C_n^b} = \infty$ for some $a$, $b$, $0 < a,b < \infty$ .

An algebra is an $F^{\mu}P^{\mu}$-algebra if it is both an $F^{\mu}$- and a $P^{\mu}$-algebra. It is simply called the F- , P- , FP-algebra for the $F^{\mu}$, $P^{\mu}$, $F^{\mu}P^{\mu}$-algebra, respectively, in case $\mu$ is the Haar measure on $\Gamma$ .

By the definitions, we have the following going up property for $P^{\mu}$-algebras and the going down property for $F^{\mu}$-algebras:

4.6. THEOREM Let $(A(G),\|\ \|_A)$ and $(B(G),\|\ \|_B)$ be two Banach algebras in $L^1(G)$ such that $A(G)$ is a subalgebra of $B(G)$ . (I) if $B(G)$ is an $F^{\mu}$-algebra, then $A(G)$ is an $F^{\mu}$-algebra. (II) if $A(G)$ is a $P^{\mu}$-algebra, then $B(G)$ is a $P^{\mu}$-algebra.

It shall be seen that many Segal algebras are F-algebras while the improper Segal algebra $L^1(G)$ is known not. Moreover;

4.7. THEOREM For any non-discrete $G$ , there exists a proper Segal algebra which is not an F-algebra.

PROOF. Take two compact symmetric neighborhoods $Q$ and $\Delta$ of $0$ such that

$$Q \subset \text{Int } \Delta$$
$$d(Q) \geq 1$$

where Int and d denote the interior and the Haar measure, respectively. As in the proof of Theorem 4.18, there exists a sequence $\Omega = (\gamma_n)$ in $\Gamma$ without cluster point and a nonzero element $\gamma_0$ in $\Gamma$ such that

$$\Omega \cap (\gamma_0 + \Omega) = \phi$$
$$\gamma - \beta \notin \Delta^2 \quad \text{for} \quad \gamma, \beta \in \Omega \cup (\gamma_0 + \Omega), \quad \gamma \neq \beta .$$

In particular, $(\gamma_n + \Delta) \cap (\gamma_m + \Delta) = \phi$ for $n \neq m$. Choose $Q_n$ and $\Delta_n$ as the union of $n^n$ number of $\gamma_i + Q$ and $\gamma_i + \Delta$, respectively, such that

$$Q_n \subset \Delta_n$$
$$\Delta_n \cap \Delta_m = \phi \quad \text{if} \quad n \neq m .$$

Since $Q_n$ and $\Delta_n$ are compact neighborhoods of 0 such that $Q_n \subset \Delta_n$, by Theorem 1.11 (ii), there is $f_n$ in $L^1(G)$ such that

$$\|f_n\|_{L^1} \leq 2$$
$$\hat{f}_n(Q_n) = 1$$
$$\text{Supp } \hat{f}_n \subset \Delta_n .$$

Now

$$\sum_{n=1}^{\infty} \frac{\|f_n\|_{L^1}}{n^2} < \infty$$

there is f in $L^1(G)$ such that

$$f = \sum_{n=1}^{\infty} \frac{f_n}{n^2} .$$

This implies

$$\hat{f} = \sum_{n=1}^{\infty} \frac{\hat{f}_n}{n^2}$$

and

$$|\hat{f}|^p = \sum_{n=1}^{\infty} \left|\frac{\hat{f}_n}{n^2}\right|^p , \qquad 0 < p < \infty .$$

Thus

$$\begin{aligned}\int_\Gamma |\hat{f}|^p \, d\gamma &= \sum_{n=1}^{\infty} \int_\Gamma \frac{|\hat{f}_n|^p}{n^{2p}} \, d\gamma \\ &\geq \sum_{n=1}^{\infty} \int_{Q_n} \frac{|\hat{f}_n|^p}{n^{2p}} \, d\gamma \\ &= \sum_{n=1}^{\infty} \frac{d(Q_n)}{n^{2p}} \\ &\geq \sum_{n=1}^{\infty} \frac{n^n}{n^{2p}} \\ &= \infty \qquad 0 < p < \infty .\end{aligned}$$

Equivalently, $\hat{f} \notin L^p(\Gamma)$ for $0 < p < \infty$ . But if we associate with $(\gamma_0 + \gamma_n)$ a discrete measure $\mu$ , $\mu(\gamma_0 + \gamma_n) = 1$ , $n = 1,2,\cdots$ , then $\mu$ is regular and hence $L^{1(1,\mu)}(G)$ is a Segal algebra. Routine arguments reveal that $f \in L^{1(1,\mu)}(G)$ and that $L^{1(1,\mu)}(G)$ is a <u>proper</u> Segal algebra. //

On the contrary, we have

4.8. <u>THEOREM</u> There is a proper Segal algebra $S(T)$ whose

Fourier transforms is contained in $\ell^p(Z)$ , for all $p$ , $0 < p \leq \infty$ .

<u>PROOF.</u> It suffices to take a $f \in C^\infty(T)$ with $\hat{f}(n) \neq 0$ for all $n$ , and then consider the Segal algebra

$$f * L^1(T) . \qquad //$$

Incidentally, we found that the class of P-algebras is large:

4.9. <u>THEOREM</u> [Wang (1972)] Every character Segal algebra is a P-algebra.

<u>PROOF.</u> Since $G$ is non-discrete, $\Gamma$ non-compact. There exists a sequence $(\gamma_n)_{n=1}^\infty$ in $\Gamma$ and a compact symmetric neighborhood $\Delta$ of $0$ such that $\gamma_1 = 0$ , $(\gamma_i + \Delta) \cap (\gamma_j + \Delta) = \phi$ if $i \neq j$ . Let $\Omega$ be a compact symmetric neighborhood of $0$ with $\Omega \subset \text{Int } \Delta$ . Let $f$ be a generalized trapezium function in $L^1(G)$ such that

$$0 \leq \hat{f} \leq 1$$
$$\hat{f}(\Omega) = 1$$
$$\text{Supp } \hat{f} \subset \Delta .$$

Then $f \in S(G)$ by Theorem 4.2 (i) . Let $\Delta_n = \gamma_n + \Delta$ , $\Omega_n = \gamma_n + \Omega$, $f_n = \gamma_n f$ , and $C_n = qn$ where $q > 1$ and $\|f\|_S \leq q$ $(n = 1,2,\cdots)$ . Then

P-1. Holds by the construction of the $\Delta$, $\Omega$, and $\gamma_n$ .

P-2. $0 \leq \hat{f}_n \leq 1$ , Supp $\hat{f}_n \subset \gamma_n + \Delta = \Delta_n$ , $\hat{f}_n(\gamma_n + \Omega) = \hat{f}_n(\Omega_n) = 1$ $(n = 1,2,\cdots)$ .

P-3. $\sum_{n=1}^{\infty} \frac{1}{(qn)^2} = \sum_{n=1}^{\infty} \frac{1}{q^2 n^2} < \infty$ , $\sum_{n=1}^{\infty} \frac{1}{qn} = \infty$ , and

$$\|f_n\|_S = \|\gamma_n f\|_S = \|f\|_S$$

$$\leq q \leq qn = C_n \quad (n = 1,2,\cdots) .$$

Thus $S(G)$ has the property P . This completes the proof.

4.10. THEOREM [Šilov] Let $S(T)$ be a Segal algebra on $T$ containing $C^{\infty}(T)$ . Then there exists a positive integer $k$ such that $S(T) \supset C^k(T)$ .

PROOF. By Theorem 3.8, for $f \in S(T)$ , we have

$$\lim_{n\to\infty} \|\hat{f}(n)e^{int}\|_S = 0 .$$

We clain that there exists a positive integer $m_0$ and constant $C$ such that

$$\|e^{int}\|_S \leq C|n|^{m_0} \quad \text{for each } n .$$

For otherwise, for $m = 1,2,\cdots$ there exists a $n_m$ such that

$$\|e^{in_m t}\|_S > |n_m|^m .$$

Consider the function $f$ defined by

$$f(t) = \sum_{m=1}^{\infty} \frac{e^{in_m t}}{|n_m|^m} .$$

Then $f \in C^{\infty}(T) \subset S(T)$ . But

$$\lim_{m\to\infty} \|\hat{f}(n_m)e^{in_m t}\|_S = \lim_{m\to\infty} \frac{\|e^{in_m t}\|_S}{|n_m|^m}$$

$$\geq 1 .$$

A contradiction. Let $k = m_o + 2$ , and consider $C^k(T)$ . It is well-known that for $f \in C^k(T)$ , we have

$$|\hat{f}(n)| \leq \frac{C_1}{|n|^k}$$

for some constant $C_1$ independent of $f$ . Furthermore

$$\sum_{n=-\infty}^{\infty} \|\hat{f}(n)e^{int}\|_S = \sum_{n=-\infty}^{\infty} |\hat{f}(n)|\ \|e^{int}\|_S$$

$$\leq \sum_{n=-\infty}^{\infty} C_1 \frac{\|e^{int}\|_S}{|n|^k}$$

$$= \sum_{n=-\infty}^{\infty} C_1 \frac{\|e^{int}\|_S}{|n|^{m_o + 2}}$$

$$\leq \sum_{n=-\infty}^{\infty} C_1 C \frac{1}{|n|^2}$$

$$< \infty .$$

Thus there exists $g \in S(T)$ such that

$$g = \sum_{n=-\infty}^{\infty} \hat{f}(n)e^{int} .$$

Therefore $\hat{g}(n) = \hat{f}(n)$ for all $n$ , or $g = f$ . We conclude that $f \in S(T)$ . This proves our assertion. //

4.11. <u>COROLLARY</u> A Segal algebra $S(T)$ , containing $C^{\infty}(T)$ , is a P-algebra.

PROOF. By Theorem 4.10 and that $C^k(T)$ , $0 \le k < \infty$ , are P-algebras. //

In what follows, we list and examine a number of natually occuring group algebras which are either Segal algebras or FP-algebras:

## 4.12. EXAMPLES

$C^k(T)$ [non-character Segal algebra and FP-algebra] For an integer $k$ , $1 \le k < \infty$ , and the circle group $T$ , $C^k(T)$ denotes the Banach algebra of all functions with $k$ continuous dervatives on $T$ under the norm

$$\|f\| = \sup_{0 \le j \le k} \|f^{(j)}\|_\infty .$$

$L^{(k)}(T)$ [non-character Segal algebra and FP-algebra] For an integer $k$ , $1 \le k < \infty$. $L^{(k)}(T)$ denotes the Banach algebra of all functions $f$ such that $f^{(j)}$ , $j = 0,1,\cdots,k-1$, are absolutely continuous and are in $L^1(T)$ under the norm

$$\|f\| = \sup_{0 \le j \le k} \|f^{(j)}\|_{L^1} .$$

BV(T) [semi-homogeneous Banach algebra and FP-algebra ] BV(T) denotes the Banach algebra of all functions $f$ which are continuous on $T$ with bounded total variation $V_o^{2\pi} f$ under the norm

$$\|f\| = \|f\|_{L^1} + V_o^{2\pi} f .$$

$Lip_\alpha(T)$ [semi-homogeneous Banach algebra and FP-algebra] For $0 < \alpha \leq 1$, $Lip_\alpha(T)$ denotes the Banach algebra of all functions $f$ of Lipschitz class $\alpha$.

D(T) [Segal algebra and P-algebra] D(T) denotes the Banach algebra of all functions $f$ in $L^1(T)$ with $\|f - D_N * f\|_{L^1} \to 0$ where $(D_N)$ is the Dirichlet kernel under the norm

$$\|f\| = \sup_{n \geq 1} \|D_n * f\|_{L^1} .$$

E(T) [semi-homogeneous Banach algebra and P-algebra] E(T) denotes the Banach algebra of all functions $f$ in $L^1(T)$ for which $\|f\| = \sup_{n \geq 1} \|D_n * f\|_{L^1} < \infty$ under the norm $\|f\|$.

C(G) [character Segal algebra and FP-algebra] For an infinite compact abelian group, C(G) denotes the Banach algebra of all continuous functions on G under the supremum norm.

$L^p(G)$ [character Segal algebra and FP-algebra] For $1 \leq p < \infty$ and G an infinite compact abelian group with normalized Haar measure $dx$, $L^p(G)$ denotes the Banach algebra of all measurable functions $f$ on G with

$$\|f\| = \left( \int_G |f(x)|^p \, dx \right)^{\frac{1}{p}} < \infty$$

under the norm $\|f\|$.

$L^\infty(G)$ [semi-homogeneous Banach algebra and FP-algebra] For an infinite compact abelian group G, $L^\infty(G)$ denotes the Banach algebra of all essentially bounded functions on G under the essential supremum norm.

$L^{(k)}(R)$ [non-character Segal algebra and FP-algebra] For an integer $k$, $1 \leq k < \infty$ and $R$ the additive group of all real numbers, $L^{(k)}(R)$ denotes the Banach algebra of all functions $f$ in $L^1(R)$ such that $f^{(j)}$, $j = 0,1,\cdots,k-1$, are absolutely continuous on $R$ and are in $L^1(R)$ under the norm

$$\|f\| = \sup_{0 \leq j \leq k} \|f^{(j)}\|_{L^1} .$$

<u>PROOF</u>. We only prove this example for the case of $k = 1$. It could be shown similarly for the other cases. It is clear that $(L^{(1)}(R), \| \|)$ is a normed linear space with $\| \|_{L^{(1)}} \geq \| \|_{L^1}$. Moreover,

I. $(L^{(1)}(R), \| \|_{L^{(1)}})$ is a Banach space.

Let $(f_n)_{n=1}^{\infty}$ be a Cauchy sequence in $(L^{(1)}(R), \| \|_{L^{(1)}})$. Then $(f_n)_{n=1}^{\infty}$ and $(f'_n)_{n=1}^{\infty}$ are Cauchy sequence in $(L^1, \| \|_{L^1})$. There exist $f,g$ in $L^1(R)$ such that $\|f_n - f\|_{L^1} \to 0$ and $\|f'_n - g\|_{L^1} \to 0$. Let $(f_{n_k})_{k=1}^{\infty}$ be a subsequence of $(f_n)_{n=1}^{\infty}$ such that $f_{n_k} \to f$ a.e. Since $\|f'_{n_k} - g\|_{L^1} \to 0$ we have $\lim_{n\to\infty} \int_{-\infty}^{x} f'_{n_k}(t)dt = \int_{-\infty}^{x} g(t)dt$ for $x \in R$. But $f_{n_k}(-\infty) = 0$ since $f_{n_k}$ is both in $L^1(R)$ and of bounded variation on $R$, and $f_{n_k}(x) = \int_{-\infty}^{x} f'_{n_k}(t)dt$. Thus

$$\begin{aligned} f(x) &= \lim_{k\to\infty} f_{n_k}(x) \qquad \text{a.e.} \\ &= \lim_{k\to\infty} \int_{-\infty}^{x} f'_{n_k}(t)dt \qquad \text{a.e.} \\ &= \int_{-\infty}^{x} g(t)dt \qquad \text{a.e.} \end{aligned}$$

Differentiating, we have $f' = g$ a.e. This proves that $\|f_n - f\|_{L^{(1)}} \to 0$ .

II. $(L^{(1)}(R), \|\ \|_{L^{(1)}})$ satisfies the properties H1 and H2.

For every $f \in L^{(1)}(R)$ and for every $a \in R$ , we have

$$(L_a f)'(x) = L_a f'(x) \qquad \text{a.e.}$$

So

$$\begin{aligned}\|L_a f\|_{L^{(1)}} &= \max\{ \|L_a f\|_{L^1} , \|(L_a f)'\|_{L^1} \} \\ &= \max\{ \|f\|_{L^1} , \|L_a f'\|_{L^1} \} \\ &= \max\{ \|f\|_{L^1} , \|f'\|_{L^1} \} \\ &= \|f\|_{L^{(1)}} .\end{aligned}$$

Therefore $L_a f \in L^{(1)}(R)$ and $\|L_a f\|_{L^{(1)}} = \|f\|_{L^{(1)}}$ .

Since $x \to L_x f$ is a continuous mapping of $G$ into $(L^1(R), \|\ \|_{L^1})$ and $(L_a f)' = L_a f'$ a.e., for every $f \in L^{(1)}(R)$, it follows that $x \to L_x f$ is a continuous mapping of $G$ into $(L^{(1)}(R), \|\ \|_{L^{(1)}})$ .

III. $(L^{(1)}, \|\ \|_{L^{(1)}})$ is a Segal algebra.

It suffices to show that $L^{(1)}(R)$ is dense in $(L^1(R), \|\ \|_{L^1})$ . Let $A(R)$ be the linear space of all infinitely differentiable functions on $R$ , which satisfy $\lim_{|x|\to\infty} x^n f^{(j)}(x) = 0$ $(n \geq 0 , j \geq 0)$ . Then we have $A(R) \subset L^{(1)}(R)$ . But $A(R)$ is dense in $(L^1(R), \|\ \|_{L^1})$ (see Katznelson (1968,p.150) so $L^{(1)}(R)$ is dense in $(L^1(R), \|\ \|_{L^1})$ .

IV. $(L^{(1)}(R), \|\ \|_{L^{(1)}})$ is non-character.

For every character $e^{ixt}$ , and every $f(t) \in L^{(1)}(R)$, we have $e^{ixt}f(t) \varepsilon L^{(1)}(R)$ . Let $f \neq 0$ in $L^{(1)}(R)$ . Then

$$\|e^{ixt}f\|_{L^{(1)}} = \max\{\|f\|_{L^1}, \|(e^{ixt}f'(t) + ixe^{ixt}f(t)\|_{L^1}\}$$
$$\geq \max\{\|f\|_{L^1}, |\ |x|\ \|f\|_{L^1} - \|f'\|_{L^1}|\}$$
$$> \max\{\|f\|_{L^1}, \|f'\|_{L^1}\}$$
$$\text{if } x \text{ satisfies } |x|\|f\|_{L^1} - \|f'\|_{L^1} > \|f'\|_{L^1}$$
$$= \|f\|_{L^{(1)}} .$$

Combining I, II, III, and IV, we completes the proof.

$L_{BV}(R)$ [semi-homogeneous Banach algebra and FP-algebra ]
$L_{BV}(R)$ denotes the Banach algebra of all functions $f$ in $L^1(R)$ which are bounded variation on $R$ under the norm

$$\|f\| = \|f\|_{L^1} + V_R f .$$

F(R) [Segal algebra and non-F algebra] F(R) denotes the Banach algebra of all functions $f$ in $L^1(R)$ with $\lim_{n\to\infty} \hat{f}(n)\log n = 0$ under the norm

$$\|f\| = \|f\|_{L^1} + \sup_{n\geq 1} |\hat{f}(n)|\log n .$$

W(G) [character Segal algebra and FP-algebra] For a non-discrete locally compact abelian group $G$ having a discrete subgroup $H$ such that $G/H$ is compact. (There will then exist a compact set $K$ of measure 1 in $G$ such that $G = HK$). For example $G = R$ , $H$ = integers, $K = [0,2\pi]$ , Haar measure = $\frac{1}{2\pi}$ . Lebesgue measure W(G) , called Wiener algebra, denotes the Banach algebra of all continuous functions $f$ on $G$ with

$$\|f\| = \sup_{u \in G} \sum_{h \in H} \max_{x \in K} |f(u + h + x)| < \infty$$

under the norm $\|f\|$.

$W_\infty(G)$ [semi-homogeneous Banach algebra and FP-algebra] For a $G$ as in $W(G)$ above, $W_\infty(G)$ denotes the Banach algebra of all functions $f$ in $L^\infty(G)$ with

$$\|f\| = \sup_{u \in G} \sum_{h \in H} \operatorname{ess\,sup}_{x \in K} |f(u + h + x)| < \infty$$

under the norm $\|f\|$.

$L^1 \cap C_0(G)$ [character Segal algebra and FP-algebra] For a non-discrete locally compact abelian group $G$, $L^1 \cap C_0(G)$ denotes the Banach algebra of all continuous functions $f$ in $L^1(G)$ which vanish at $\infty$ under the norm

$$\|f\| = \|f\|_{L^1} + \|f\|_\infty .$$

$L^1 \cap L^p(G)$ [character Segal algebra and FP-algebra] For a non-discrete locally compact abelian group $G$, $L^1 \cap L^p(G)$ denotes the Banach algebra of all functions $f$ in $L^1(G)$ and $L^p(G)$ under the norm

$$\|f\| = \|f\|_{L^1} + \|f\|_{L^p} .$$

$L^1 \cap L^\infty(G)$ [semi-homogeneous Banach algebra and FP-algebra] For a non-discrete locally compact abelian group $G$, $L^1 \cap L^\infty(G)$ denotes the functions $f$ in $L^1(G)$ and $L^\infty(G)$

under the norm

$$\|f\| = \|f\|_{L^1} + \|f\|_\infty .$$

$S^p(\mu)$ [Segal algebra and $F^\mu P^\mu$-algebra] For $1 \leq p < \infty$ , and $G$ a non-discrete locally compact abelian group with character group $\Gamma$ and $\mu$ a positive unbounded regular measure on $\Gamma$ . Let $S(G)$ be a Segal algebra. $S^p(\mu)$ denotes the Banach algebra of all functions $f$ in $S(G)$ such that $\hat{f} \in L^p(\mu)$ under the norm

$$\|f\| = \|f\|_S + \|\hat{f}\|_{L^p(\mu)}$$

$S^p(\mu)$ forms a Segal and $F^\mu P^\mu$-algebra. $S^p(\mu)$ is character if $S(G)$ is character and $\mu$ is translation-invariant.

PROOF. By Theorem 3.13-7-1, $S^{(p,\mu)}(G)$ forms a homogeneous Banach algebra. To see that $S^{(p,\mu)}(G)$ is a Segal algebra, it suffices to show that $P(L^1(G))$ is contained in $S^{(p,\mu)}(G)$ . For $f \in P(L^1(G))$ , let $K = \text{Supp } \hat{f}$ . Then $K$ is compact, and hence $\mu(K) < \infty$ since $\mu$ is regular. Therefore $\hat{f} \in L^p(\mu)$ . Finally we claim that $S^{(p,\mu)}(G)$ is character if $S(G)$ is character and $\mu$ is translation-invariant. In fact, for $f \in S^{(p,\mu)}(G)$ , $\gamma \in \Gamma$ , then

$$\begin{aligned} \|\gamma f\| &= \|\gamma f\|_S + \|\widehat{\gamma f}\|_{L^p(\mu)} \\ &= \|f\|_S + \|L_\gamma \hat{f}\|_{L^p(\mu)} \\ &= \|f\|_S + \|\hat{f}\|_{L^p(\mu)} \\ &= \|f\| . \end{aligned}$$

//

As a consequence, $A^p(G)$ is a character Segal algebra.

$\mu * S(G)$ [Segal algebra] For $\mu \in M(G)$ whose Fourier-Stieltjes transform $\hat{\mu}$ is never vanished and $S(G)$ a Segal algebra. $\mu * S(G)$ denotes the Banach algebra of all functions $\mu * f$, where $f \in S(G)$ under the norm

$$\|\mu * f\| = \|f\|_S .$$

$S_f$ [Segal algebra] For $f \in L^1(G)$ whose Fourier transform $\hat{f}$ is never vanished, and $S(G)$ a Segal algebra. $S_f$ denotes the Banach algebra of all functions $g$ in $S(G)$ with $f * g \in C_o(G)$ under the norm

$$\|g\| = \|g\|_{L^1} + \|f * g\|_\infty .$$

$S_1^{(\xi,S_2)}(G)$ [Segal algebra] Let $G$ be a non-discrete locally compact abelian group, and $\xi$ a continuous function on $\Gamma$ such that $\xi\widehat{P(L^1(G))} \subset \widehat{P(L^1(G))}$. Such a $\xi$ exists, for instance, take $\xi = \hat{\mu}$ for some $\mu \in M(G)$, or take $\xi$ as any function on $\Gamma$ whenever $G$ is compact. Let $S_1(G)$, and $S_2(G)$ be two Segal algebras, and let $S_1^{(\xi,S_2)}(G)$ be the space of all functions $f$ in $S_1(G)$ such that $\xi\hat{f} \in S_2(G)$. Under the sum norm $\|f\| = \|f\|_{S_1} + \|\sigma(f)\|_{S_2}$, where $\sigma(f) \in S_2(G)$ and $\widehat{\sigma(f)} = \xi\hat{f}$, $S_1^{(\xi,S_2)}(G)$ forms a Segal algebra.

PROOF. By Theorem 3.13-8-1, $S_1^{(\xi,S_2)}(G)$ is a homogeneous Banach algebra. For $f \varepsilon P(L^1(G))$, then, by the hypothesis, $\xi\hat{f}$ is a Fourier transform of some function $g$ in $P(L^1(G))$. $P(L^1(G))$ is contained in $S_2(G)$, so $P(L^1(G)) \subset S_1^{(\xi,S_2)}(G)$. This asserts that $S_1^{(\xi,S_2)}(G)$ is a Segal algebra. //

<u>More Segal algebras</u> will be produced by the following result.

4.13. <u>THEOREM</u> Let $S(G)$ be a Segal algebra. Then a subset is a closed invariant subspace of $S(G)$ if and only if it is a closed ideal of $S(G)$.

<u>PROOF.</u> By Theorems 3.13-9-1, 3.13-9-2 and 4.4 (iii). //

4.14. <u>THEOREM</u> The algebras described above form a couple of chains:

I. Let $T$ be the circle group, then

$$\begin{array}{ccccccccccccccccccc}
C^\infty(T) & \subset \cdots \subset & C^k(T) & \subset \cdots \subset & C^1(T) & \subset & Lip_1(T) & \subset & A^1(T) & \subset \cdots \subset & A^2(T) & \subset \cdots \subset & L^1(T) & & & & & & \\
 & & \cap & & \cap & & & & \cap & & \| & & & & & & & & \\
C^\infty(T) & \subset \cdots \subset & L^{(k)}(T) & \subset \cdots \subset & L^{(1)}(T) & \subset & BV(T) & \subset & C(T) & \subset & L^\infty(T) & \subset \cdots \subset & L^2(T) & \subset \cdots \subset & L^1(T) & & & &
\end{array}$$

II. For a suitable $G$ (such as $G = R$), we have

$$\begin{array}{ccccccc}
 & & L_{BV}(G) & \subset & W_\infty(G) & \subset & L^1 \cap L^\infty(G) \\
 & & \cup & & \cup & & \\
\cdots \subset L^{(k)}(G) \subset \cdots & \subset & L^{(1)}(G) & \subset & W(G) & \subset & L^1 \cap C_o(G) \subset L^1 \cap L^\infty(G)
\end{array}$$

and $L^1 \cap L^\infty(G) \subset L^1 \cap L^p(G) \subset A^t(G) \subset L^1(G)$

where $p > 1$ ,

$$t = \begin{cases} 2 & \text{if} \quad p > 2 \\ \frac{p}{p-1} & \text{if} \quad p \leq 2 . \end{cases}$$

III. For two Segal algebras $S_1(G) \subset S_2(G)$ , we have

$$S_1^p(\mu) \subset S_1^s(\mu) \subset S_2^s(\mu) \subset S_2(G) \qquad (1 \leq p < s < \infty)$$

$$S_1(G) = S_1^{(\hat{\mu},S_1)}(G) \subset S_2^{(\hat{\mu},S_1)}(G) \subset S_2^{(\hat{\mu},S_2)}(G)$$

$$= S_2(G) \qquad (\mu \in M(G)) .$$

<u>PROOF.</u> First we shall prove that $L^{(1)}(R) \subset W(R)$ . Let $f \in L^{(1)}(R)$ and $V_{-n}^{n}$ the total variation on $[-n,n]$ . By the Radon-Nikodym Theorem, $V_R f = \|f'\|_{L^1}$ . Moreover,

$$\sum_{k=-n}^{n-1} \max_{x\varepsilon[0,1]} |f(k+x)| \leq \sum_{k=-n}^{n-1} \min_{x\varepsilon[0,1]} |f(k+x)| + V_{-n}^{n} f$$

$$\leq \|f\|_{L^1} + \|f'\|_{L^1}$$

$$= 2\|f\|_{L^{(1)}} .$$

Or

$$\sum_{k=-\infty}^{\infty} \max_{x\varepsilon[0,1]} |f(k+x)| \leq 2\|f\|_{L^{(1)}} .$$

But

$$\sup_{u\varepsilon R} \sum_{n=-\infty}^{\infty} \max_{x\varepsilon[0,1]} |g(u+n+x)| \leq 2 \sum_{n=-\infty}^{\infty} \max_{x\varepsilon[0,1]} |g(n+x)| .$$

Therefore

$$\|f\|_W \leq 4\|f\|_{L^{(1)}} .$$

Similarly it may be shown that $L_{BV}(R) \subset W_\infty(R)$ .

The other inclusions among the chains are clear.

4.15. REMARK If we examine the chains I and II above, we'll find that each of them, except the improper one $L^1$, is a F-algebra, by the going down property. Moreover each of chain I is a P-algebra by the going up property and Corollary 4.11. Using the same reason together with that $L^{(k)}(R)$ is a P-algebra and W(G) is a character Segal algebra, each of chain II is a P-algebra.

4.16. THEROEM $L^{(k)}(R)$ is a P-algebra.

PROOF. $L^{(k)}(R)$ satisfies the property P. Let $\Delta_n = [n-\frac{1}{4}, n+\frac{1}{4}]$, and $\Omega_n = [n-\frac{1}{8}, n+\frac{1}{8}]$ for $n = 1,2,\cdots$. There exists a generalized trapezium function f in $L^1(R)$ such that $0 \leq \hat{f} \leq 1$, Supp $\hat{f} \subset \Delta_1$, and $\hat{f}(\Omega_1) = 1$. Then $f \in L^{(k)}(R)$ since it is a Segal algebra. Let $f_n(t) = e^{i(n-1)t}f(t)$, and $C_n = 2^k(n-1)^k\|f\|_{L^{(k)}} \geq 1$ for $n = 2,3,\cdots$

$$f^{(j)}(t) = (e^{i(n-1)t})^{(j)}f(t) + \binom{j}{1}(e^{i(n-1)t})^{(j-1)}f^{(1)}(t)$$

$$+ \binom{j}{2}(e^{i(n-1)t})^{(j-2)}f^{(2)}(t) + \cdots + e^{i(n-1)t}f^{(j)}(t)$$

$$= i^j(n-1)^je^{i(n-1)t}f(t) + \binom{j}{1}i^{j-1}(n-1)^{j-1}e^{i(n-1)t}f^{(1)}(t)$$

$$+ \binom{j}{2}i^{n-2}(n-1)^{n-2}e^{i(n-1)t}f^{(2)}(t)+\cdots+e^{i(n-1)t}f^{(j)}(t) .$$

Or

$$\|f_n^{(j)}\|_{L^1} \leq (n-1)^j\|f\|_{L^{(k)}} \left[1 + \binom{j}{1} + \binom{j}{2} + \cdots + \binom{j}{j}\right]$$

$$= 2^j(n-1)^j\|f\|_{L^{(k)}} .$$

Hence

$$\|f_n\|_{L^{(k)}} = \max_{j=0,1,\cdots,k} \|f_n^{(j)}\|_{L^1}$$

$$\leq 2^k(n-1)^k \|f\|_{L^{(k)}} .$$

Obviously,

$$\sum \frac{1}{C_n^{k+1}} < \infty \quad \text{but} \quad \sum \frac{1}{C_n^{\frac{1}{k}}} = \infty .$$

Then it follows that $L^{(k)}(R)$ satisfies the property $P$ . //

4.17. REMARK If we give a further examination, we find

I. $S^p(\mu)$ is an $F^{\mu}$-algebra and $A^p(G)$ is an FP-algebra.

II. If $S_1(G)$ and $S_2(G)$ are character, then $S_1^{(1,S_2)}(G)$ is character, and hence an P-algebra.

III. $S_1^{(\xi,S_2)}(G)$ is unnecessary an F-algebra since $S(G) = S^{(1,S)}(G)$

By comparing Segal algebras with the improper Segal algebra, though they share many property while they are quite diverse. Reiter (1968,p.131) and (1971,p.26) asked if there are any Segal algebras which are unstable under the multipliction by characters or are unstable under involution. The conjectures are positive as follows:

4.18. THEOREM Let $G$ be non-discrete. Then there exists a Segal algebra on $G$ which is not stable under the multiplication by characters.

PROOF. By the hypothesis, $\Gamma$ is non-compact. Take an infinite sequence $(\gamma_n)_{n=1}^{\infty}$ in $\Gamma$ without any cluster point and $\gamma_i \neq \gamma_j$

if $i \neq j$ . (see Theorem 1.2 (iv)) . This implies that $(\gamma_n)$ is not contained in any compact subset of $\Gamma$ . Then

(i) For any compact symmetric neighborhood $\Delta$ of $0$ , there exists a subsequence $(\gamma_{n_i})$ of $(\gamma_n)$ such that $\gamma_{n_i} - \gamma_{n_j} \notin \Delta$ for $i \neq j$ .

Take $\gamma_{n_1} = \gamma_1$ , and if $\gamma_{n_1}, \cdots, \gamma_{n_i}$ have been chosen, then take $\gamma_{n_{i+1}} \notin \bigcup_{i=1}^{i} (\gamma_{n_j} + \Delta)$ . Clearly $(\gamma_{n_i})$ is a candidate.

(ii) For any nonzero $\gamma_0$ in $\Gamma$ , there exists a subsequence $(\gamma_{n_i})$ of $(\gamma_n)$ such that $\Omega \cap (\gamma_0 + \Omega) = \phi$ where $\Omega = (\gamma_{n_i})$ .

Take $\gamma_{n_1} = \gamma_1$ ($\gamma_{n_1} = \gamma_2$ if necessary), and if $\gamma_{n_1}, \cdots, \gamma_{n_i}$ have been chosen, then take $\gamma_{n_{i+1}} \notin \{ \gamma_0, \gamma_{n_1}, \cdots, \gamma_{n_i}, \pm\gamma_0 + \gamma_{n_1}, \cdots, \pm\gamma_0 + \gamma_{n_i} \}$ . Routine arguments reveal that $(\gamma_{n_i})$ is a candidate.

By (i) and (ii), we may assume $(\gamma_n)$ providing the following property: For a nonzero $\gamma_0$ in $\Gamma$ and for a compact subset $\Delta$ of $\Gamma$ , we have

$$\Omega \cap (\gamma_0 + \Omega) = \phi$$

$$\gamma - \beta \notin \Delta \quad \text{for} \quad \gamma \neq \beta \ , \ \gamma, \beta \in \Omega \cup (\gamma_0 + \Omega)$$

where $\Omega = (\gamma_n)$ . Take a generalized triangle function $\sigma$ in $L^1(G)$ such that

$$\sigma \geq 0$$

$$\| \sigma \|_{L^1} = \hat{\sigma}(0) = 1$$

$$\text{Supp } \hat{\sigma} \subset \Delta .$$

Since

$$\sum_{n=1}^{\infty} \frac{\| (\gamma_0 + \gamma_n)\sigma \|_{L^1}}{n^2} < \infty .$$

there is $f$ in $L^1(G)$ such that

$$f = \sum_{n=1}^{\infty} \frac{(\gamma_0 + \gamma_n)\sigma}{n^2} .$$

Let $\mu$ be a discrete measure associated with $(\gamma_n)$ defined by

$$\mu(\gamma_n) = n ,$$

$\mu$ is regular since $(\gamma_n)$ has no cluster points. Then consider the Segal algebra $L^{1(1,\mu)}(G)$ , we have

$$\sum_{n=1}^{\infty} \hat{f}(\gamma_n)\mu(\gamma_n) = 0$$

and

$$\sum_{n=1}^{\infty} \widehat{(-\gamma_0)f}(\gamma_n)\mu(\gamma_n) = \sum_{n=1}^{\infty} \hat{f}(\gamma_0 + \gamma_n)\mu(\gamma_n)$$

$$= \sum_{n=1}^{\infty} \frac{1}{n}$$

$$= \infty .$$

This asserts that $f \in L^{1(1,\mu)}(G)$ but $(-\gamma_0)f \notin L^{1(1,\mu)}(G)$ . //

4.19. <u>THEOREM</u> There exists a Segal algebra which is not stable under the involution.

<u>PROOF.</u> Recall that for any $f$ in $L^1(G)$ , $f^*$ , the involution of $f$ , is defined by

$$f^*(x) = \overline{f(-x)} .$$

Then

$$\hat{f}^*(\gamma) = \overline{\hat{f}(\gamma)} .$$

Let $T$ be the circle group, and

$$f(x) = \sum_{n=2}^{\infty} \frac{\cos nx}{\sqrt{n} \log n}$$

$$= \sum_{|n|\geq 2} \frac{1}{2\sqrt{|n|} \log |n|} e^{inx} .$$

Then, by Zygmund (1959, p.125, p.187), we have

(i) $f \in L^2(T)$

(ii) $f \notin L^p(T)$ for $p > 2$

(iii) There is a sequence $(\varepsilon_n = \pm 1)$ such that

$$\sum_{n=2}^{\infty} \varepsilon_n \frac{\cos nx}{\sqrt{n} \log n} \in L^p(T) \quad \text{for} \quad p > 0 .$$

Let

$$\xi(n) = \begin{cases} \sqrt{\varepsilon_{|n|}} & \text{if} \quad |n| \geq 2 \\ 0 & \text{if} \quad |n| < 2 . \end{cases}$$

We have

$$\sum_{n=2}^{\infty} \xi(n) \frac{\cos nx}{\sqrt{n} \log n} \in L^{2(\xi, L^3)}(T)$$

but

$$\sum_{n=2}^{\infty} \overline{\xi(n)} \frac{\cos nx}{\sqrt{n} \log n} \notin L^{2(\xi, L^3)}(T) .$$

Therefore $L^{2(\xi,L^3)}(T)$ is a Segal algebra which is not stable under the involution. //

Burnham (1972,[1]), (1975,[4]) generalize the (classical) Segal algebras to the A-(abstract) Segal algebras:

**4.20. DEFINITION** Let $(A, \| \ \|_A)$ be a Banach algebra. The proper subalgebra B of A is called an A-Segal algebra if

A1. B is a dense left ideal of A .

A2. $(B, \| \ \|_B)$ is a Banach algebra.

A3. There exists $M > 0$ such that

$$\|f\|_A \leq M\|f\|_B \quad (f \in B) .$$

A4. There exists $C > 0$ such that

$$\|fg\|_B \leq C\|f\|_A\|g\|_B .$$

Abstract Segal algebra share some properties with (classical) Segal algebras. Burnham (1972,[1]) asserts that if A, B are commutative Banach algebras and B is an A-Segal algebra, then they have the same ideal theory. However, A-Segal algebra may posses a left identity (see Leinert (1975)) while a Segal algebra never even admit a left bounded approximate identity.

## 5. BEURLING ALGEBRAS

The main purpose of this section is to establish several theorems from which we'll see how great a difference there is among Beurling algebras, homogeneous Banach algebras, and FP-algebras. For instance, a proper Beurling algebra is not an ideal in $L^1(G)$ whereas the others are.

5.1. <u>DEFINITIONS</u> An upper semi-continuous real-valued function $w$ on $G$ is said to be a weight function if it satisfies the following properties:

(i) $1 \leq w(x) \quad (x \in G)$

(ii) $w(x + y) \leq w(x)w(y) \quad (x,y \in G)$

(iii) $w$ is locally bounded in the sense of that $w$ is bounded on every compact set in $G$.

For any weight function $w$ on $G$, let $L_w^1(G)$ be the subalgebra of all functions $f$ in $L^1(G)$ with $fw \in L^1(G)$. Under the norm

$$||f||_w = ||fw||_{L^1} ,$$

$L_w^1(G)$ forms a Banach algebra. We say that $L_w^1(G)$ is the <u>Beurling algebra defined by $w$.</u> Clearly, if $w$ is bounded (in particular, if $G$ is compact), then $L_w^1(G) = L^1(G)$. $L^1(G)$ is the largest Beurling algebra. A Beurling algebra is said to be <u>proper</u> if it is not the whole of $L^1(G)$. Define

$$w_1(t) = e^{a|t|} , \quad a \geq 0$$

and

$$w_2(t) = \begin{cases} e^t & \text{for } t \geq 0 \\ 1 & \text{otherwise .} \end{cases}$$

Then $w_1$ and $w_2$ are weight functions on $R$. Moreover, $w_3$ is a weight function on $R^n$, where

$$w_3(t) = (1 + |t|)^a , \quad a \geq 0$$

5.2. THEOREM Let $L^1_w(G)$ be a Beurling algebra. Then

(i) $C_c(G)$ is dense in $L^1_w(G)$ .

(ii) For $f \varepsilon L^1_w(G)$ , $x \varepsilon G$ , then $f_x \varepsilon L^1_w(G)$ , and $\|f_x\|_w \leq w(x)\|f\|_w$ .

(iii) $L^1_w(G)$ satisfies H2, that is, $x \to f_x$ is continuous for any $f$ in $L^1_w(G)$ .

PROOF.

(i) Given $f \varepsilon L^1_w(G)$ , and $\varepsilon > 0$ , take $k_1 \varepsilon C_c(G)$ such that $\|fw - k_1\|_{L^1} < \frac{\varepsilon}{2}$ . Let $K_1$ be a compact set whose interior contains Supp $k_1$ . Then $w(x) \leq A$ , say, if $x \varepsilon K_1$ . There is a $k \varepsilon C_c(G)$ such that Supp $k \subset K_1$ , and $\|\frac{k_1}{w} - k\|_{L^1} < \frac{\varepsilon}{2A}$ . Then $\|f - k\|_w = \|(f - k)w\|_{L^1} < \varepsilon$ .

(ii) For $f \varepsilon L^1_w(G)$ , $x \varepsilon G$ , then

$$\begin{aligned} \int |f_x(y)|w(y)dy &= \int |f(y - x)|w(y)dy \\ &= \int |f(y)|w(x + y)dy \\ &\leq \int |f(y)|w(x)w(y)dy \\ &= w(x)\|f\|_w < \infty . \end{aligned}$$

Therefore $f_x \varepsilon L^1_w(G)$ , and $\|f_x\|_w \leq w(x)\|f\|_w$ .

(iii) Since $\|f_x\|_w \leq w(x)\|f\|_w$ , it suffices to show that $x \to f_x$ is continuous at $0$ . Let $W$ be a compact neighborhood of $0$ , then $w(x) \leq A$ , say, for $x \varepsilon W$ . Given $f \varepsilon L^1_w(G)$ , and $\varepsilon > 0$ , take $k \varepsilon C_c(G)$ such that

$\|f - k\|_w < \frac{\varepsilon}{2(A+1)}$ . There is a neighborhood U of 0 contained in W such that $\|k_y - k\|_w < \frac{\varepsilon}{2}$ for $y \varepsilon U$ . Then

$$\begin{aligned} \|f_y - f\|_w &\leq \|f_y - k_y\|_w + \|k_y - k\|_w + \|k - f\|_w \\ &\leq (w(y) + 1)\|f - k\|_w + \|k_y - k\|_w \\ &< (A + 1)\frac{\varepsilon}{2(A+1)} + \frac{\varepsilon}{2} \\ &= \varepsilon \quad (y \varepsilon U) . \end{aligned}$$ //

As in the $L^1(G)$ , it can be shown that $L^1_w(G)$ admits an identity if and only if G is discrete. Moreover,

5.3. <u>THEOREM</u> A Beurling algebra $L^1_w(G)$ admits a bounded approximate identity. In particular, $L^1_w(G)$ admits the factorization property.

<u>PROOF</u>. Let K be a compact neighborhood of 0 , then $w(x) \leq A$, say, for $x \varepsilon K$ . Let $\mathfrak{F}$ be the family of all neighborhoods of 0 contained in K . For $U, V \varepsilon \mathfrak{F}$ , defined $U \geq V$ if $U \subset V$ . Then clearly $(\mathfrak{F}, \geq)$ is a directed set. For every $U \varepsilon \mathfrak{F}$ , there exists a positive continuous function $h_U$ on G such that $\int h_U(x)dx = 1$ and the support of $h_U$ is contained in U . Consider the net $(h_U)_{U \in \mathfrak{F}}$ , we have $\|h_U\|_w \leq A$ for each $U \varepsilon \mathfrak{F}$ . By Theroem 5.2 (iii), given $f \varepsilon L^1_w(G)$ and $\varepsilon > 0$ , take W in $\mathfrak{F}$ such that $x \varepsilon W$ implies $\|L_x f - f\|_w < \varepsilon$ .

For every $U \varepsilon \mathfrak{F}$ with $U \geq W$ , we have

$$\begin{aligned}\|h_U * f - f\|_w &= \int |h_U * f(x) - f(x)|w(x)dx \\ &= \int \left| \int h_U(y)f(x-y)dy - \int h_U(y)f(x)dy \right| w(x)dx \\ &\leq \int |f(x-y) - f(x)|h_U(y)dy \int w(x)dx \\ &= \int |f(x-y) - f(x)|w(x)dx \int h_U(y)dy \\ &= \int \|L_y f - f\|_w h_U(y)dy < \varepsilon \int h_U(y)dy \\ &= \varepsilon .\end{aligned}$$

Therefore, $(h_U)_{U \in \mathfrak{F}}$ is a bounded approximate identity for $(L_w^1(G), \|\ \|_w)$. The factorization of $L_w^1(G)$ follows from Theorem 8.3. //

5.4. THEOREM Let $L_{w_1}^1(G)$ and $L_{w_2}^1(G)$ be two Beurling algebras with $L_{w_1}^1 \subset L_{w_2}^1(G)$, then $L_{w_1}^1(G) * L_{w_2}^1(G) = L_{w_2}^1(G)$. In particular, $L_{w_1}^1(G)$ is an ideal in $L_{w_2}^1(G)$ if and only if $L_{w_1}^1(G) = L_{w_2}^1(G)$.

PROOF. In order to prove that $L_{w_1}^1(G) * L_{w_2}^1(G) = L_{w_2}^1(G)$, it suffices to prove that $L_{w_2}^1(G)$ is a Banach $L_{w_1}^1(G)$-module, and that $L_{w_1}^1(G) * L_{w_2}^1(G)$ is dense in $L_{w_2}^1(G)$. Clearly $L_{w_2}^1(G)$ is a module over $L_{w_1}^1(G)$. Moreover, by Theorem 1.14(iii), there exists a constant $C \geq 1$ such that $\|\ \|_{w_2} \leq C\|\ \|_{w_1}$. For $f \in L_{w_1}^1(G)$, $g \in L_{w_2}^1(G)$, we have

$$\|f * g\|_{w_2} \leq \|f\|_{w_2} \|g\|_{w_2}$$
$$\leq C\|f\|_{w_1} \|g\|_{w_2} .$$

Hence $L^1_{w_2}(G)$ is a Banach $L^1_{w_1}(G)$-module. Moreover, consider the following inclusions:

$$C_c(G) \subset L^1_{w_1}(G) = L^1_{w_1}(G) * L^1_{w_1}(G) \subset L^1_{w_1}(G) * L^1_{w_2}(G) \subset L^1_{w_2}(G) .$$

We obtain that $L^1_{w_1}(G) * L^1_{w_2}(G)$ is dense in $L^1_{w_2}(G)$ since $C_c(G)$ is dense in $L^1_{w_2}(G)$. This completes the proof. //

5.5. COROLLARY If a homogeneous Banach algebra is a Beurling algebra, then it is the whole of $L^1(G)$.

PROOF. By Theorems 3.3 and 5.4. //

As we have just seen, a homogeneous Banach algebra is an ideal in the $L^1$-algebra but a proper Beurling algebra is only a subalgebra of the $L^1$-algebra. Another difference between these two algebras appeares on their maximal ideal spaces. Let $L^1_w(G)$ be a Beurling algebra. Then the maximal ideal space $m(L^1_w(G))$ of $L^1_w(G)$ contains the maximal ideal space $m(L^1(G))$, which is $\Gamma$, as a closed subspace. This is due to that $L^1_w(G)$ is dense in $L^1(G)$ [Loomis (1953,Theorem 24B)]. It reveals that the maximal ideal space of a Beurling algebra is larger than that of $L^1(G)$. But the maximal ideal space of a homogeneous Banach algebra is smaller than that of $L^1(G)$; Theorem 11.3.

In order to consider the maximal ideal space of a Beurling algebra, we review the following facts.

Let $L_w^1(G)$ be a Beurling algebra and let $L_w^\infty(G)$ be the algebra of all measurable functions $\eta$ on $G$ for which

$$\|\eta\|_{\infty,w} = \operatorname*{ess\,sup}_{x \in G} \left|\frac{\eta(x)}{w(x)}\right| < \infty .$$

Under the $\| \ \|_{\infty,w}$ , $L_w^\infty(G)$ forms a Banach algebra, which is also the dual space of $L_w^1(G)$ [Reiter (1968,p.84)]. More precisely, for $\eta \in L_w^\infty(G)$ , $\sigma : f \to \int f(x)\overline{\eta(x)}dx$ is a continuous linear functional on $L_w^1(G)$ with $\|\sigma\| = \|\eta\|_{\infty,w}$ , and every continuous linear functional on $L_w^1(G)$ is of the form above.

A generalized character of $G$ is a continuous function $\gamma$ on $G$ such that $|\gamma(x)| \neq 0$ and $\gamma(x + y) = \gamma(x)\gamma(y)$ for $x,y \in G$ . Clearly a character of $G$ is a generalized character.

5.6. <u>THEOREM</u> Let $L_w^1(G)$ be a Beurling algebra, and $\eta$ a generalized character with $|\eta(x)| \le w(x)$ locally almost everywhere (l.a.e.). Then the map $\sigma_\eta$ defined by

$$\sigma_\eta(f) = \int f(x)\overline{\eta(x)}dx$$

is a complex homomorphism of $L_w^1(G)$ . Moreover, every complex homomorphism of $L_w^1(G)$ is obtained in this way, and if $\sigma_{\eta_1} = \sigma_{\eta_2}$ , then $\eta_1 = \eta_2$ .

<u>PROOF</u>. Let $\eta$ be a generalized character with $|\eta(x)| \le w(x)$ l.a.e., and $\sigma_\eta$ as in the theorem. Then obviously $\sigma_\eta$ is a

non-trivial linear functional on $L_w^1(G)$ since $\eta \in L_w^\infty(G)$. Moreover, for $f,g \in L_w^1(G)$,

$$\begin{aligned}
\sigma_\eta(f * g) &= \int (f * g)(x)\overline{\eta(x)}dx \\
&= \int f(x - y)g(y)dy \int \overline{\eta(x)}dx \\
&= \int f(x - y)\overline{\eta(x)}dx \int g(y)dy \\
&= \int f(x)\overline{\eta(x + y)}dx \int g(y)dy \\
&= \int f(x)\overline{\eta(x)}dx \int g(y)\overline{\eta(y)}dy \\
&= \sigma_\eta(f)\sigma_\eta(g) .
\end{aligned}$$

Therefore $\sigma_\eta$ is a complex homomorphism of $L_w^1(G)$.

Conversely suppose that $\sigma$ is a complex homomorphism of $L_w^1(G)$, then $\sigma$ is a bounded linear functional on $L_w^1(G)$ with $||\sigma|| \leq 1$. Take $\eta \in L_w^\infty(G)$ such that

$$\sigma(f) = \int f(x)\overline{\eta(x)}dx .$$

For $f,g \in L_w^1(G)$,

$$\begin{aligned}
\int \sigma(f)g(y)\overline{\eta(y)}dy &= \sigma(f)\sigma(g) = \sigma(f * g) \\
&= \iint f(x - y)g(y)\overline{\eta(x)}dxdy \\
&= \int g(y)\sigma(f_y)dy
\end{aligned}$$

or, $\sigma(f)\overline{\eta(y)} = \sigma(f_y)$ l.a.e. Take $f$ such that $\sigma(f) \neq 0$. Since $y \to \sigma(f_y)$ is continuous, we may assume that $\eta$ is continuous and $\eta(0) \neq 0$ . Let $x,y \in G$ , then

$$\sigma(f)\overline{\eta(x+y)} = \sigma(f_{x+y}) = \sigma(L_y f_x)$$
$$= \sigma(f_x)\overline{\eta(y)}$$
$$= \sigma(f)\overline{\eta(x)}\,\overline{\eta(y)}$$

for all $f$ , so $\eta(x+y) = \eta(x)\eta(y)$ , and hence $\eta(x) \neq 0$ for $x \in G$ . This asserts that $\eta$ is a generalized character with $|\eta(x)| \leq w(x)$ l.a.e., and $\sigma = \sigma_\eta$ .

Finally if $\sigma_{\eta_1} = \sigma_{\eta_2}$ , then clearly $\eta_1 = \eta_2$ . //

As a consequence of the last theorem, the maximal ideal space of a Beurling algebra $L^1_w(G)$ could be identified with the space of all generalized characters $\eta$ with $\eta \in L^\infty_w(G)$ . For instance, for $a > 0$ , the maximal ideal space of a Beurling algebra $L^1_w(G)$ defined by $w(t) = e^{a|t|}$ could be identified with the space of all complex members $c + id$ such that $|c| \leq a$ , which contains the real line as a closed (identified) subspace.

In the final of this section, we wish to mention some facts about the ideal theory of Beurling algebras. The following Lemma will be useful.

**5.7.** <u>LEMMA</u> [Porcelli (1966,p.88)]

(i) Let $A$ be a Banach algebra with $A^2 \neq 0$ , then $A^2 \neq A$ if and only if $A$ contains a non-prime maximal ideal.

(ii) Let $A$ be a Banach algebra without identity and $I$

a maximal ideal in $A$ , then $I$ is regular if and only if $I$ is prime.

5.8. <u>COROLLARY</u> A maximal ideal in $L_w^1(G)$ is regular and closed.

<u>PROOF</u>. It is well-known that a regular maximal ideal in a Banach algebra is closed. Therefore, it suffices to show that a maximal ideal $M$ in $L_w^1(G)$ is regular. Suppose that $L_w^1(G)$ admits an identity, then it is done. If $L_w^1(G)$ does not admit an identity, and $M$ is a maximal ideal in $L_w^1(G)$ , then $M$ is regular by Lemma 5.7 and $(L_w^1(G))^2 = L_w^1(G)$ . //

It needs one more Lemma to prove our theorem.

5.9. <u>LEMMA</u> [Dietrich (1972)] If $I_1$ and $I_2$ are ideals in an algebra $A$ such that $I_1 \subset I_2$ and there is no ideal of $A$ strictly between them, then $AI_2 \subset I_1$ or $MI_2 \subset I_1$ for some maximal ideal $M$ in $A$ containing $I_1$ .

By a Beurling algebra $L_w^1(G)$ satisfying the condition $D$ , we mean that for every $\eta$ in the maximal ideal space of $L_w^1(G)$, there exists a net $(f_\lambda)$ in $L_w^1(G)$ such that

$$\hat{f}_\lambda = 0 \text{ near } \eta \text{ for each } \lambda$$

and

$$f_\lambda * f \to f$$

for every $f$ in $L_w^1(G)$ such that $\hat{f}(\eta) = 0$ .

5.10. <u>THEOREM</u> Let $L_w^1(G)$ be a Beurling algebra satisfying the condition D. Suppose that $I_1$ and $I_2$ are two ideals in $L_w^1(G)$ such that $I_1$ is closed, and $I_1 \subsetneqq I_2$, then there is an ideal of $L_w^1(G)$ strictly between $I_1$ and $I_2$.

<u>PROOF</u>. If there are no ideals strictly between $I_1$ and $I_2$, then by Lemma 5.9, $L_w^1(G) * I_2 \subset I_1$, or $M * I_2 \subset I_1$ for some maximal ideal in $L_w^1(G)$. That M is regular follows from Theorem 5.8. Choose $\eta$ in the maximal ideal space of $L_w^1(G)$ such that

$$M = \{f \in L_w^1(G) : \hat{f}(\eta) = 0\} .$$

Then M admits an approximate identity since $L_w^1(G)$ satisfies the condition D. We then have that both $L_w^1(G)$ and M admit an approximate identity. But this implies

$$I_2 \subset \overline{I_w^1(G) * I_2}^{L_w^1} .$$

Or

$$I_2 \subset \overline{M * I_2}^{L_w^1} .$$

It turns out

$$I_2 \subset I_1 \subset I_2$$

in any cases since $I_1$ is closed. This is a contradiction. Therefore there is an ideal strictly between $I_1$ and $I_2$. //

# 6. TILDE ALGEBRAS

6.1. <u>DEFINITION</u> Let $A$ be a Banach algebra and $B$ a Banach algebra in $A$, we define the <u>relative completion</u> $\tilde{B}^A$ of $B$ w.r.t. $A$ to be

$$\tilde{B}^A = \bigcup_{\eta>0} \overline{S_B(\eta)}^A$$

where $S_B(\eta) = \{f \in B : \|f\|_B \leq \eta\}$ and $\overline{E}^A$ is the closure of $E$ in the A-norm. For each $f \in \tilde{B}^A$, we set

$$\|f\|_{\tilde{B}^A} = \inf \{t : f \in \overline{S_B(t)}^A\} .$$

It is easy to see that $(\tilde{B}^A, \|\ \|_{\tilde{B}^A})$ is a Banach algebra and that $B$ is a subalgebra of $\tilde{B}^A$ with $\|f\|_{\tilde{B}^A} \leq \|f\|_B$ for all $f \in B$.

Let $E$ be a closed subset of $T$. Consider $A(E)$ as a Banach algebra in $C(E)$. Then $\tilde{A}(E) \supset A(E)$ and $\|f\|_{\tilde{A}(E)} \leq \|f\|_{A(E)}$ for all $f \in A(E)$. What is the relation between $A(E)$ and $\tilde{A}(E)$?

Katznelson-McGehee (1968, 1971) has shown that there is $A(E)$ with

$$A(E) \neq \tilde{A}(E) \quad \text{and} \quad \|f\|_{A(E)} = \|f\|_{\tilde{A}(E)} \quad \text{for all} \quad f \in A(E) .$$

Varopoulos (1971,Sect. 2, Chap. 12) proved that we can have $A(E) = \tilde{A}(E)$ yet

$$\sup_{0\neq f\in A(E)} \frac{\|f\|_{A(E)}}{\|f\|_{\tilde{A}(E)}} \geq 1 + \varepsilon .$$

Moreover, Katznelson-Körner (1973) proved that

(i) There is a closed set $E$ such that $A(E) \neq \tilde{A}(E)$ but $A(E)$ is dense in $(\tilde{A}(E), \|\ \|_{\tilde{A}(E)})$.

(ii) For every $K \geq 1$ there exists a closed set $E$ such that $A(E) = \tilde{A}(E)$ but for some $0 \neq f \in A(E)$,

$$\|f\|_{A(E)} \geq K\|f\|_{\tilde{A}(E)} .$$

There is another story for the Segal algebras case: Burnham (1975,[4]) established many interesting results:

6.2. LEMMA Let $B$ be an A-Segal algebra. If $f \in \tilde{B}^A$, then there exists a sequence $(f_n)$ in $B$ such that

(i) $\|f_n - f\|_A \to 0$

(ii) $\|f_n\|_B \to |||f|||$ .

PROOF. Let $f \in \tilde{B}^A$ and $a_k = |||f||| + \frac{1}{k}$, $k = 1,2,\cdots$, then $f \in \overline{S_B(a_k)}^A$. Thus for each $k$, there is $f_n \in S_B(a_n)$ with $\|f_n - f\|_A < \frac{1}{n}$, $n = 1,2,\cdots$. Hence $\|f_n - f\|_A \to 0$ as $n \to \infty$.

Since $\|f_n\|_B \leq a_n$,

$$\limsup_{n \to \infty} \|f_n\|_B \leq |||f||| .$$

Suppose $\liminf_{n \to \infty} \|f_n\|_B < \delta < |||f|||$, then there exists some subsequence of $(f_n)$ which lies in $S_B(\delta)$, or $f \in \overline{S_B(\delta)}^A$ contradicting $\delta < |||f|||$. Thus $|||f||| \leq \liminf_{n \to \infty} \|f_n\|_B$. Therefore

$$\|f_n\|_B \to |||f||| .$$

Let $E$ be a common right approximate identity of $B$ and $A$ with $\|e\|_A = 1$ for all $e \in E$.

6.3. THEOREM If $B$ is an A-Segal algebra, then the following two conditions are equivalent:

(i) $f \in \tilde{B}^A$

(ii) $M = \sup_{e \in E} \|fe\|_B < \infty$.

If either condition holds then $M = |||f|||$.

PROOF. Suppose $f \in \tilde{B}^A$. Then there exists a sequence $(f_n)$ in $B$ so that

$$\|f_n - f\|_A \to 0 \qquad \|f_n\|_B \to |||f||| .$$

Then, for each $e \in E$, we have

$$\|f_n e - fe\|_B \leq \|f_n - f\|_A \|e\|_B \to 0 .$$

Hence

$$\|fe\|_B = \lim_{n\to\infty} \|f_n e\|_B \leq \lim_{n\to\infty} \|f_n\|_B$$

$$= |||f||| .$$

Thus (ii) holds and $M \leq |||f|||$.

Now suppose (ii) holds for some $f \in A$. Choose $(e_n) \subset E$ with $\|fe_n - f\|_A \to 0$. By the hypothesis $fe_n \in S_B(M)$. Hence $f \in \tilde{B}^A$ with $|||f||| \leq M$. Hence (i) holds. But then $M \leq |||f|||$ so that $M = |||f|||$.

6.4. THEOREM If $B$ is an A-Segal algebra, then $\tilde{B}^A$ is an A-Segal algebra. Furthermore, $B$ is a closed ideal of $\tilde{B}^A$ with

$\|f\|_B = |||f|||$ for all $f \in B$ .

PROOF. Routine arguments reveal that $\tilde{B}^A$ is an A-Segal algebra and $B$ is an ideal of $\tilde{B}^A$ .

It suffices to show that $\|f\|_B \leq |||f|||$ for all $f \in B$ . If $f \in B$ , then $f \in \tilde{B}^A$ so that

$$\|fe\|_B \leq |||f||| \quad \text{for} \quad e \in E .$$

Thus

$$\|f\|_B \leq |||f||| \qquad (f \in B) .$$

Therefore

$$\|f\|_B = |||f||| \qquad (f \in B) .$$

6.5. DEFINITION We say that an A-Segal algebra $B$ is singular if $B$ can be embedded as a closed (left, right, two-sided) ideal in some A-Segal algebra $B'$ with $B \subsetneqq B' \subsetneqq A$ . Segal algebras that are not singular are called non-singular.

6.6. THEOREM If $B$ is an A-Segal algebra, then $B$ is singular if and only if $B \subsetneqq \tilde{B}^A$ .

PROOF. Since $\tilde{B}^A \subsetneqq A$ , it suffices to show that the only-if-part. Suppose $B$ is singular and $B'$ is as specified in above Definition. Let $f \in B'$ . Let $(e_n) \subset B$ with

$$\|e_n\|_A \leq 1 , \qquad \|fe - f\|_A \to 0 .$$

So

$$\|fe_n\|_{B'} \leq \|f\|_{B'} .$$

But $B$ is closed in $B'$ so there exists $M > 0$ with $\|g\|_B \leq M\|g\|_{B'}$ $(g \in B)$. Take $g = fe_n$, we have

$$\|fe_n\|_B \leq M\|fe_n\|_{B'} \leq M\|f\|_{B'} .$$

Thus $f \in \tilde{B}^A$, or $B' \subseteq \tilde{B}^A$. Thus

$$B \subsetneq \tilde{B}^A .$$

6.7. <u>COROLLARY</u> If $B'$ is an A-Segal algebras which contains $B$ as a closed ideal, then $B' \subseteq \tilde{B}^A$. In particular, if $\tilde{B}^A \subseteq B'$ then $B' = \tilde{B}^A$.

Using the Module Factorization Theorem one can prove

6.8. <u>THEOREM</u> (i) If $f \in \tilde{B}^A$, then $f \in B \Longleftrightarrow$ given any $\varepsilon > 0$ there exists $e(f,\varepsilon) = e$ in a common right approximate identity with $\|e\|_A = 1$ and $|||fe - f||| < \varepsilon$. Furthermore, if $B'$ is any A-Segal algebra for which $AB' = B$, then $B' \subseteq \tilde{B}^A$.

(ii) Let $f \in \tilde{S}$. Then $f \in S$ if and only if $|||f_a - f||| \to 0$ as $a \to 0$.

Burnham's theory on the relative completion has been generalized by Lakien (1975) and Feichtinger (1976):

6.9. <u>DEFINITION</u> Let $B$ be a Segal algebra on the locally compact abelian group $G$. We let $\tilde{B}$ be the set of $\mu \in M(G)$ such that there exists a net $(f_\alpha) \subseteq B$ with

$$\mu = (\text{weak*}) \lim f_\alpha$$

and

$$\sup_{\alpha} \|f_\alpha\|_B < \infty .$$

Let $\|\mu\|_{\tilde{B}}$ be the $\inf_{f_\alpha} \sup_{\alpha} \|f_\alpha\|_B$. We shall see that $(\tilde{B}, \|\ \|_{\tilde{B}})$ forms a Banach algebra which is called the w-relative completion of $B$.

6.10. <u>THEOREM</u> If $\mu \in \tilde{B}$, then $\|\mu\|_{\tilde{B}} \geq \|\mu\|_M$.

<u>PROOF</u>. Given any $\varepsilon > 0$, there exists a net $(f_\alpha) \subseteq B$ with

$$f_\alpha \xrightarrow{\omega^*} \mu \quad \text{and} \quad \sup_{\alpha} \|f_\alpha\|_B \leq \|\mu\|_{\tilde{B}} + \varepsilon .$$

Hence

$$\|\mu\|_M = \sup_{\substack{g\in C_o(G) \\ \|g\|_\infty \leq 1}} |\langle g,\mu\rangle| = \sup_{\substack{g\in C_o(G) \\ \|g\|_\infty \leq 1}} \lim_{\alpha} |\langle g,f_\alpha\rangle|$$

$$\leq \sup_{\substack{g\in C_o(G) \\ \|g\|_\infty \leq 1}} (\sup \|f_\alpha\|_B)\|g\|_\infty \leq \sup_{\alpha} \|f_\alpha\|_B$$

$$\leq \|\mu\|_{\tilde{B}} + \varepsilon .$$

Since $\varepsilon$ is arbitrary, $\|\mu\|_M \leq \|\mu\|_{\tilde{B}}$. //

6.11. <u>THEOREM</u> $(\tilde{B}, \|\ \|_{\tilde{B}})$ forms a Banach algebra.

<u>PROOF</u>. Let's prove that

(i) Since $\|\ \|_{\tilde{B}} \geq \|\ \|_M$, $\|\mu\|_{\tilde{B}} = 0 \Longrightarrow \mu = 0$.

(ii) $\|\mu + \nu\|_{\tilde{B}} \leq \|\mu\|_{\tilde{B}} + \|\nu\|_{\tilde{B}}$ for $\mu, \nu \in \tilde{B}$.

Let $(f_\alpha)_{\alpha\in D_1}$, $(g_\alpha)_{\alpha\in D_2}$ be two nets in $B$ satisfying

$$f_\alpha \xrightarrow{\omega^*} \mu ,$$

$$\sup_\alpha \|f_\alpha\|_B \le \|\mu\|_{\tilde{B}} + \frac{\varepsilon}{2} .$$

$$g_\alpha \xrightarrow{\omega^*} \nu ,$$

$$\sup_\alpha \|g_\alpha\|_B \le \|\nu\|_{\tilde{B}} + \frac{\varepsilon}{2} .$$

Consider the directed set $D = D_1 \times D_2$ and define

$$f'_\alpha = f_{\alpha_1}$$
$$g'_\alpha = g_{\alpha_2} \qquad \text{for} \quad \alpha = (\alpha_1, \alpha_2) .$$

We have

$$f'_\alpha \xrightarrow{\omega^*} \mu ,$$

$$\sup_\alpha \|f'_\alpha\|_B \le \|\mu\|_{\tilde{B}} + \frac{\varepsilon}{2} .$$

$$g'_\alpha \xrightarrow{\omega^*} \nu ,$$

$$\sup_\alpha \|g'_\alpha\|_B \le \|\nu\|_{\tilde{B}} + \frac{\varepsilon}{2} .$$

Hence

$$\|\mu + \nu\|_{\tilde{B}} \le \sup_\alpha \|f'_\alpha + g'_\alpha\|_B \le \sup_\alpha \|f'_\alpha\|_B + \sup_\alpha \|g'_\alpha\|_B$$
$$\le \|\mu\|_{\tilde{B}} + \|\nu\|_{\tilde{B}} + \varepsilon .$$

Since $\varepsilon$ is arbitrary, $\|\mu + \nu\|_{\tilde{B}} \le \|\mu\|_{\tilde{B}} + \|\nu\|_{\tilde{B}}$ .

(iii) For $\mu, \nu \in \tilde{B}$ , we have $\mu * \nu \in \tilde{B}$ with $\|\mu * \nu\|_{\tilde{B}} \le \|\mu\|_{\tilde{B}} \|\nu\|_{\tilde{B}}$. Take $(f_\alpha) \subseteq B$ , $\sup_\alpha \|f_\alpha\|_B \le \|\mu\|_{\tilde{B}} + \varepsilon/(\|\nu\|_{\tilde{B}} + 1)$ and $f_\alpha \xrightarrow{\omega^*} \mu$ . We have

$$f_\alpha * \nu \xrightarrow{\omega^*} \mu * \nu ,$$

$f_\alpha * \nu \in B$ and $\sup_\alpha \|f_\alpha * \nu\|_B \le \sup_\alpha \|f_\alpha\|_B \|\nu\|_M$

$\le \|\mu\|_{\tilde{B}} \|\nu\|_{\tilde{B}} + \varepsilon$ , so $\mu * \nu \in \tilde{B}$ and $\|\mu * \nu\|_{\tilde{B}} \le \|\mu\|_{\tilde{B}} \|\nu\|_{\tilde{B}}$.

(iv) $(\tilde{B}, \| \ \|_{\tilde{B}})$ is complete.

Suppose that $(\mu_n)$ is a Cauchy sequence in $\tilde{B}$. Without loss of generality, we can assume that $\|\mu_n - \mu_{n+1}\|_{\tilde{B}} < \frac{1}{2^n}$, and $\mu_n \xrightarrow{\| \ \|_M} \mu$ for some $\mu \in M(G)$. We need to prove $\mu_n \xrightarrow{\| \ \|_{\tilde{B}}} \mu$. By the definition of $\mu_n$, there exist countable nets $(f_{n\alpha}) \subseteq B$ satisfying the following properties:

$$\sup_\alpha \|f_{1\alpha}\|_B < \infty , \qquad f_{1\alpha} \xrightarrow{\omega^*} \mu_1$$

and for $n \ge 2$,

$$\sup_\alpha \|f_{n\alpha}\|_B \le \frac{1}{2^{n-1}} , \qquad f_{n\alpha} \xrightarrow{\omega^*} \mu_n - \mu_{n-1} .$$

As in the proof of part (ii), we can assume that all the nets $(f_{n\alpha})$ have the same index set. Define $g_{n\alpha} = \sum_{k=1}^{n} f_{n\alpha}$, we have

$$\sup_\alpha \|g_{1\alpha}\|_B < \infty , \qquad \sup_\alpha \|g_{n\alpha} - g_{(n-1)\alpha}\|_B \le \frac{1}{2^{n-1}}$$

and

$$g_{n\alpha} \xrightarrow{\omega^*} \mu_n \quad \text{as} \quad \alpha \to \infty .$$

Therefore there exist $g_\alpha \in B$ such that $g_{n\alpha}$ converges to $g_\alpha$ uniformly on $\alpha$ under the $\| \ \|_B$-norm. From the following diagram,

$$\begin{array}{ccc} g_{n\alpha} & \xrightarrow{\omega^*} & \mu_n \\ \| \ \|_B \downarrow & & \downarrow \| \ \|_M \\ g_\alpha & & \mu \end{array}$$

it is easy to see that $g_\alpha \xrightarrow{\omega^*} \mu$ and

$$\|g_\alpha\|_B = \lim_{n\to\infty} \|g_{n\alpha}\|_B \leq \|g_{1\alpha}\|_B + \frac{1}{2} + \frac{1}{2^2} + \cdots + \frac{1}{2^n} + \cdots$$

$$= \|g_{1\alpha}\|_B + 1 ,$$

so $\mu \in \tilde{B}$ and

$$\|\mu_n - \mu\|_{\tilde{B}} \leq \sup_\alpha \|g_{n\alpha} - g_\alpha\|_B \leq \frac{1}{2^{n-1}} \to 0 \qquad \text{as} \quad n \to \infty .$$

Thus $\mu_n \xrightarrow{\| \ \|_{\tilde{B}}} \mu$ .

(i), (ii), (iii), (iv) together with some other trivial properties reveal that $(\tilde{B}, \| \ \|_{\tilde{B}})$ forms a Banach algebra. //

6.12. <u>THEOREM</u> $\tilde{B}$ is an ideal in $M(G)$ with

$$\|\mu * \nu\|_{\tilde{B}} \leq \|\mu\|_M \|\nu\|_{\tilde{B}} \qquad (\mu \in M, \ \nu \in \tilde{B}) .$$

<u>PROOF</u>. For $\mu \in M(G)$ , $\nu \in \tilde{B}$ , take $(f_\alpha) \subseteq B$ with $\sup_\alpha \|f_\alpha\|_B \leq c$ , $f_\alpha \xrightarrow{\omega^*} \nu$ . Then $(\mu * f_\alpha) \subseteq B$ ,

$$\sup_\alpha \|\mu * f_\alpha\|_B \leq \sup_\alpha \|\mu\|_M \|f_\alpha\|_B = \|\mu\|_M \sup_\alpha \|f_\alpha\|_B \quad \text{and}$$

$$\begin{aligned} \langle h, \mu * f_\alpha\rangle &= \langle h * \mu, f_\alpha\rangle \\ &\to \langle h * \mu, \nu\rangle \\ &= \langle h, \mu * \nu\rangle \qquad (h \in C_o(G)) . \end{aligned}$$

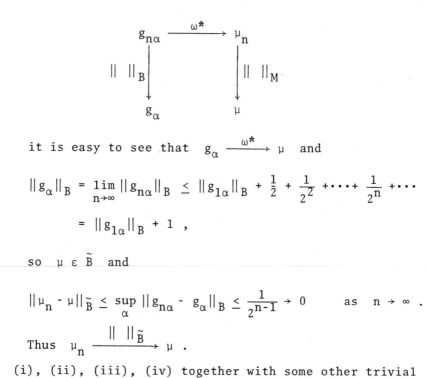

Therefore $\mu * \nu \in \tilde{B}$ and

$$\|\mu * \nu\|_{\tilde{B}} \leq \sup_{\alpha} \|\mu * f_\alpha\|_B \leq \|\mu\|_M \|\nu\|_{\tilde{B}} . \qquad //$$

Moreover, Feichtinger (1976, Theorem 3.7) established

6.13. THEOREM Let $B$ be a character Segal algebra on a non-discrete locally compact abelian group $G$. Then

$$\|f\|_B = \|f\|_{\tilde{B}} \qquad (f \in B) .$$

PROOF. First, we claim that for each $f \in L^1(G)$, $\gamma \to \gamma f$ is a continuous map of $\Gamma$ into $L^1(G)$. In fact, for $\varepsilon > 0$, take a compact set $K$ in $G$ with

$$\int_{G \setminus K} |f(x)|dx < \frac{\varepsilon}{4} .$$

Fixed $\gamma_0 \in \Gamma$, consider the neighborhood $\gamma_0 + N$ of $\gamma_0$, where $N = \{\gamma \in \Gamma : |(x,\gamma) - 1| < \frac{\varepsilon}{2(\|f\|_1+1)}$ for $x \in K\}$. For $\gamma \in \gamma_0 + N$, we have

$$\int_K |(\gamma f)(x) - (\gamma_0 f)(x)|dx < \frac{\varepsilon}{2} .$$

Or,

$$\int_G |(\gamma f)(x) - (\gamma_0 f)(x)|dx < \varepsilon .$$

Secondly, for each $f \in B$, $\gamma \to \gamma f$ is a continuous map of $\Gamma$ into $B$. It suffices to prove that $\gamma \to \gamma f$ is continuous for $f \in P(L^1(G))$, since $P(L^1(G))$ is dense in $B$.

For $f \in P(L^1(G))$, $\gamma_0 \in \Gamma$, let $K_1 = \text{supp}\,\hat{f}$. Take a compact neighborhood $K_2$ of $\gamma_0$. Put $K_3 = K_2 + K_1$. Choose a constant $C_{K_3}$ such that

$$\|g\|_B \leq C_{K_3} \|g\|_{L^1}$$

for each $g \in L^1(G)$ with supp $\hat{g} \subset K_3$ . Thus

$$\|\gamma f - \gamma_0 f\|_B \leq C_{K_3} \|\gamma f - \gamma_0 f\|_{L^1} \qquad (\gamma \in K_2) .$$

Therefore $\gamma \to \gamma f$ is a continuous map of $\Gamma$ into $B$ for each $f$ in $P(L^1(G))$ .

The previous arguments reveal that $(\hat{B}, \| \|_{\hat{B}})$ forms a homogeneous Banach space under the norm $\|\hat{f}\|_{\hat{B}} = \|f\|_B$ . By Theorem 2.11, $\hat{B}$ forms a Banach $L^1(\Gamma)$-module under

$$f \circledast \hat{g} = \int f(\gamma) L_\gamma \hat{g} d\gamma \qquad (f \in L^1(\Gamma), \ \hat{g} \in \hat{B})$$

and $L^1(\Gamma) \circledast \hat{B} = \hat{B}$ . As a matter of fact, $\circledast$ is the same as the usual convolution $*$ . Recall that $M(\Gamma) \subset (\hat{B})^*$ , for $\mu \in M(\Gamma)$ ,

$$\begin{aligned} \langle f \circledast \hat{g}, \mu\rangle &= \int \langle f(\gamma) L_\gamma \hat{g}, \mu\rangle d\gamma \\ &= \iint f(\gamma)\hat{g}(\beta-\gamma) d\mu(-\beta) d\gamma \\ &= \iint f * \hat{g}(\beta) d\mu(-\beta) \\ &= \langle f * \hat{g}, \mu\rangle \quad . \end{aligned}$$

Thus

$$f \circledast \hat{g} = f * \hat{g} \qquad (f \in L^1(\Gamma), \ \hat{f} \in \hat{B}) .$$

Clearly, $\|f\|_{\tilde{B}} \leq \|f\|_B$ for $f \in B$ . Moreover, we claim that $\|f\|_B \leq \|f\|_{\tilde{B}}$ for $f \in B$ . Since $P(L^1(G))$ is dense in $B$ , it suffices to prove that $\|f\|_B \leq \|f\|_{\tilde{B}}$ for $f \in P(L^1(G))$.

For $f \in P(L^1(G))$ , $\varepsilon > 0$ , take $\mu \in P(L^1(\Gamma))$ such that $\|\mu\|_{L^1(\Gamma)} = 1$ , $\|\hat{f} - \mu * \hat{f}\|_{\hat{B}} < \varepsilon$ . But $\widehat{\check{\mu} f} = \mu * \hat{f}$ , so $\|f - \check{\mu} f\|_B < \varepsilon$ . Choose $h \in P(L^1(G))$ , $\|h\| = 1$ , $\|\check{\mu} f - h * \check{\mu} f\|_B < \varepsilon$ . Let $(f_\alpha) \subset B$ , $\sup_\alpha \|f_\alpha\|_B \leq C$ , $f_\alpha \xrightarrow{\omega^*} f$ . We claim that $\|h * \check{\mu} f_\alpha - h * \check{\mu} f\|_B \to 0$ as $\alpha \to \infty$ . By the Module Factorization Theorem, it suffices to show that

$$\|h * \check{\mu} f_\alpha - h * \check{\mu} f\|_{L^1} \to 0 \quad \text{as} \quad \alpha \to \infty .$$

It is clear that there is, for $\varepsilon > 0$ , a compact set $K_\varepsilon$ in $G$ with

$$\int_{G \setminus K_\varepsilon} |h * \check{\mu}(f_\alpha - f)(y)|\,dy < \frac{\varepsilon}{2} \cdots\cdots\cdots\cdots \quad (1)$$

Choose an open symmetric neighborhood $U$ of $0$ such that, for $y - x \in U$ ,

$$\|L_y h - L_x h\|_\infty < \frac{\varepsilon}{4(\sup_\alpha \|f_\alpha - f\|_{L^1} + 1)|K_\varepsilon|} .$$

Choose $\alpha_x$ such that

$$|\langle L_{-x} h, \check{\mu}(f_\alpha - f)\rangle| < \frac{\varepsilon}{4|K_\varepsilon|} \qquad (\alpha \geq \alpha_x) .$$

For $y \in x + U$ , we have

$$\langle L_{-y} h, \check{\mu}(f_\alpha - f)\rangle = \langle L_{-y} h - L_{-x} h, \check{\mu}(f_\alpha - f)\rangle + \langle L_{-x} h, \check{\mu}(f_\alpha - f)\rangle .$$

Or,

$$|\langle L_{-y} h, \check{\mu}(f_\alpha - f)\rangle| < \frac{\varepsilon}{2|K_\varepsilon|} \qquad (\alpha \geq \alpha_x) .$$

Since $K_\varepsilon$ is compact, there are finite $x_1, x_2, \cdots, x_n$ in $K_\varepsilon$ such that $K_\varepsilon \subset \bigcup_{i=1}^{n} (x_i + U)$ .

Take $\alpha_0 = \max(\alpha_{x_1}, \alpha_{x_2}, \cdots, \alpha_{x_n})$ . For $\alpha \geq \alpha_0$ , we have

$$|\langle L_{-y}h, \hat{\check{\mu}}(f_\alpha - f)\rangle| < \frac{\varepsilon}{2|K_\varepsilon|} \qquad (y \in K_\varepsilon) .$$

Recall that

$$\langle L_{-y}h, \hat{\check{\mu}}(f_\alpha - f)\rangle = h * \hat{\check{\mu}}(f_\alpha - f)(y) .$$

Therefore, for $\alpha \geq \alpha_0$ ,

$$\int_{K_\varepsilon} |h * \hat{\check{\mu}}(f_\alpha - f)(y)|dy < \frac{\varepsilon}{2} \cdots\cdots\cdots\cdots\cdots (2)$$

By (1) and (2), $\displaystyle\int_G |h * \hat{\check{\mu}}(f_\alpha - f)(y)|dy < \varepsilon \qquad (\alpha \geq \alpha_0)$ .

Or,

$$||h * \hat{\check{\mu}}f_\alpha - h * \hat{\check{\mu}}f||_{L^1} \to 0 \quad \text{as} \quad \alpha \to \infty .$$

All together we have $||f - h * \hat{\check{\mu}}f_\alpha||_B < 3\varepsilon$ , $(\alpha \geq \alpha_0)$ .

That is, $h * \hat{\check{\mu}}f_\alpha \xrightarrow{B} f$ . Therefore

$$\begin{aligned} ||f||_B &= \lim_\alpha ||h * \hat{\check{\mu}}f_\alpha||_B \\ &\leq \sup_\alpha ||h||_{L^1} \; ||\hat{\check{\mu}}f_\alpha||_B \\ &\leq \sup_\alpha ||f_\alpha||_B . \end{aligned}$$

Or,

$$||f||_B \leq ||f||_{\tilde{B}} .$$

This completes the proof. //

6.14. <u>THEOREM</u> Let $B$ be a character Segal algebra on a non-discrete locally compact abelian group $G$ . Then

$$L^1 * \tilde{B} \subseteq B \quad \text{and} \quad ||f * g||_B \leq ||f||_{L^1} ||g||_{\tilde{B}} .$$

<u>PROOF</u>. Let $f \in L^1(G)$ , $g \in \tilde{B}$ . Take $(f_n) \subset P(L^1(G))$ ,

$(g_\alpha) \subset B$ with

$$\|f_n\|_{L^1} = 1 \qquad \text{for all } n$$

$$f_n * f \xrightarrow{L^1} f$$

$$\sup_\alpha \|g_\alpha\|_B \leq C$$

$$g_\alpha \xrightarrow{\omega^*} g .$$

Since

$$\|g * f_n * f - g * f\|_{\tilde{B}} = \|g * (f_n * f - f)\|_{\tilde{B}}$$
$$\leq \|g\|_{\tilde{B}} \|f_n * f - f\|_{L^1} \longrightarrow 0$$

$g * f_n * f \xrightarrow{\tilde{B}} g * f$ , $g * f_n * f$ is in $B$ , so $g * f \in B$ by Theorem 6.13. Moreover,

$$\|f * g\|_B = \|f * g\|_{\tilde{B}} \leq \|f\|_M \|g\|_{\tilde{B}} = \|f\|_{L^1} \|g\|_{\tilde{B}} . \qquad //$$

Lakien (1975) proved an interesting result:

6.15. <u>THEOREM</u> Let $B(G)$ be a character Segal algebra. Then $\tilde{B}$ has an identity if and only if $B$ has a bounded approximate identity. (i.e. $B = L^1(G)$) .

<u>PROOF</u>. Let $\tilde{B}$ has an identity $e$ . Let $(e_\alpha)$ be a common approximate identity of $L^1$ and $B$ with $\|e_\alpha\|_{L^1} \leq K$ . Then $(e_\alpha * e) \subseteq B$ , and

$$(e_\alpha * e) * f = e * (e_\alpha * f)$$
$$= e_\alpha * f$$
$$\xrightarrow{B} f$$

$$\|e_\alpha * e\|_B \le \|e_\alpha\|_{L^1}\|e\|_{\tilde{B}}$$
$$\le K\|e\|_{\tilde{B}} .$$

Thus $B$ has a bounded approximate identity.

Conversely let $B$ has a bounded approximate identity, or $B = L^1(G)$ . Let $\delta_o$ be the identity of $M(G)$ . Set $(f_n)$ be an approximate identity of $L^1(G)$ with $\|f_n\|_{L^1} = 1$ . Since $C_o(G)$ is a $L^1$-module, we have for $g \in C_o(G)$

$$\|g * f_n - g\|_\infty \to 0 \text{ as } n \to \infty$$

or

$$g * f_n(0) \to g(0) .$$

That is

$$\int g(-t)f_n(t)dt \to \int g(-t)d\delta_o(t) .$$

Thus

$$\delta_o = (\text{weak*}) \lim_n f_n .$$

This asserts that $\delta_o \in \tilde{B}$ . It is easy to see that $\delta_o$ is an identity of $\tilde{B}$ .

6.16. <u>REMARK</u> Theorems 6.13, 6.14 and 6.15 also hold for any Segal algebra on an infinite compact abelian group, see Lakien (1975).

# CHAPTER III

# FACTORIZATION AND NONFACTORIZATION

Cohen proved that any Banach algebra, with a bounded left approximate identity, admits the factorization property. Thereafter, the problems of bounded approximate identities and the factorization properties in Banach algebras have attracted a number of mathematicians. We wish to study the problems of bounded approximate identities in section 7 and those of factorization in section 8.

## 7. BOUNDED APPROXIMATE IDENTITIES

In this section we wish to study the relationships among various bounded approximate identitites in Banach algebras.

First we introduce some definitions:

7.1. <u>DEFINITIONS</u> Let $(B, \|\ \|)$ be a Banach algebra.

(1) $B$ admits a weak left approximate identity if, given $f \in B$ and $\varepsilon > 0$, there is $g \in B$ such that

$$\|gf - f\| < \varepsilon .$$

(2) $B$ admits a bounded pointwise weak left approximate identity if given $f \in B$, there is a constant $K$ depending on $f$ such that for $\varepsilon > 0$, there is $g \in B$ such that

$$\|g\| \leq K , \qquad \|gf - f\| < \varepsilon .$$

(3) B admits a bounded weak left approximate identity if B admits a weak approximate identity $g$ such that $\|g\| \leq C$, where $C$ is independent of $f \in B$ and $\varepsilon > 0$.

(4) B admits a left approximate identity if, given any finite set $\{f_1, \cdots, f_n\}$ in B and $\varepsilon > 0$, there is $g \in B$ such that

$$\|gf_i - f_i\| < \varepsilon \quad \text{for} \quad i = 1, \cdots, n .$$

(4) is equivalent to

(4') B admits a left approximate identity if there is a net $(f_\lambda)$ in B such that

$$f_\lambda f \to f \quad \text{for each} \quad f \quad \text{in} \quad B .$$

(5) B admits a bounded left approximate identity if B admits a left approximate identity $(f_\lambda)$ such that $\|f_\lambda\| \leq C$, where C is a constant.

(6) B admits a strong left approximate identity if there is a sequence $(f_n)$ in B such that

$$f_n f \to f \quad \text{for each} \quad f \quad \text{in} \quad B .$$

(7) B admits a bounded strong left approximate identity if B admits a strong left approximate identity $(f_n)$ such that $\|f_n\| \leq C$, where C is a constant.

Replacing in (1) ~ (7) the word left by right and hf

by fh , we obtain the corresponding right-hand conditions for weak right approximate identities, right approximate identities and so on. We simply call that B admits an approximate identity if B admits a left approximate identity which is also a right approximate identity. We define the others in the same way. Plainly, in B , a strong left approximate identity is a left approximate identity and a left approximate identity is a weak left approximate identity. Also a bounded strong left approximate identity is a bounded left approximate identity, which is a bounded weak left approximate identity, and a bounded weak left approximate identity is a bounded pointwise weak left approximate identity. For the converse relationship, a number of mathematicians have been concerned.

Wichmann (1973) improved the above theorem, in which Banach algebras are replaced by normed algebras. Moreover, Wichmann proved the following:

7.2. THEOREM A commutative normed algebra with bounded pointwise weak approximate identity has a weak approximate identity (possibly unbounded).

A commutative normed algebra A which does not consist entirely of topological divisors of zero has bounded pointwise weak approximate identity if and only if A has a weak bounded approximate identity.

In the same year, Liu-Rooij-Wang (1973,Lemma 12) proved:

7.3. <u>THEOREM</u> A commutative Banach algebra with bounded point-wise weak approximate identity has a bounded weak approximate identity.

7.4. <u>THEOREM</u> A Banach algebra $A$ admits a left approximate identity bounded by $K$ if and only if $A$ admits a weak left approximate identity bounded by $K$.

<u>PROOF.</u> Let $A$ be a Banach algebra with a weak left approximate identity bounded by $K$. With the convention of 1, we have, for every $x \in A$ and $\varepsilon > 0$, there exists $\mu \in A$ such that

$$\|\mu\| \leq K \quad \text{and} \quad \|(1 - \mu)x\| < \varepsilon .$$

Now let $x_1, \cdots, x_n$ be any finite set of elements in $A$. Given $\varepsilon > 0$, we can choose successively $\mu_1, \cdots, \mu_n$ in $A$ such that

$$\|\mu_i\| \leq K \quad \text{and} \quad \|(1-\mu_i) \cdots (1-\mu_2)(1-\mu_1)x_i\| < \varepsilon$$

for $i = 1, 2, \cdots, n$. Define $\upsilon$ in $A$ by

$$1 - \upsilon = (1-\mu_n)\cdots(1-\mu_2)(1-\mu_1) .$$

Then

$$\begin{aligned} \|x_i - \upsilon x_i\| &= \|[(1-\mu_n)\cdots(1-\mu_{i+1})][(1-\mu_i)\cdots(1-\mu_1)x_i]\| \\ &\leq (1 + K)^{n-i}\|(1-\mu_i)\cdots(1-\mu_1)x_i\| \\ &< (1 + K)^n \cdot \varepsilon . \end{aligned}$$

Finally we choose $\mu$ in $A$ with $\|\mu\| \leq K$, and

$\| \upsilon - \mu\upsilon\| < \varepsilon$ . Then for $i = 1,2,\cdots,n$ we have

$$\|x_i - \upsilon x_i\| \le \|x_i - \upsilon x_i\| + \|(\upsilon - \mu\upsilon)x_i\| + \|\mu(x_i - \upsilon x_i)\|$$

$$\le \|x_i - \upsilon x_i\| + \|\upsilon - \mu\upsilon\| \cdot \|x_i\| + K\,\|x_i - \upsilon x_i\|$$

$$< (1+K)^n \cdot \varepsilon + \|x_i\| \cdot \varepsilon + (1+K)^{n+1} \cdot \varepsilon .$$

Hence, for every finite set $x_1,\cdots,x_n$ of elements in $A$ and every $\varepsilon > 0$ , there exists $\mu$ in $A$ such that $\|\mu\| \le K$ and $\|x_i - \mu x_i\| < \varepsilon$ for $i = 1,2,\cdots,n$ . This concludes that $A$ admits a left approximate identity bounded by $K$ .

It is well-known that $L^1(G)$ admits a bounded approximate identity. Moreover,

7.5. THEOREM

(i) A proper Segal algebra admits an unbounded approximate identity.

(ii) For any compact abelian group $G$ , a homogeneous Banach algebra $B(G)$ on $G$ admits an approximate identity.

(iii) If $G$ is a non-compact locally compact abelian group, then there is a homogeneous Banach algebra which admits no approximate identity.

PROOF. (i) and (ii) follow from Theorems 4.4 (iii) and 3.7 (iv), respectively.

(iii) Suppose that $G$ is a non-compact locally compact abelian group, then the spectral synthesis fails for $L^1(G)$ .

Moreover, it is well-known that there is $f \in L^1(G)$ such that the closed ideals generated by $f^n$, $n = 1,2,\cdots$ are all distinct. In particular, $f$ does not belong to the closed ideal generated by $f^2$. Denote the closed ideal generated by $f$ and $f^2$ by $\langle f \rangle$ and $\langle f^2 \rangle$, respectively. Then, plainly,

$$\langle f \rangle^2 \subseteq \langle f^2 \rangle \subsetneq \langle f \rangle .$$

We conclude that $\langle f \rangle^2$ is not dense in $\langle f \rangle$. Therefore $\langle f \rangle$ admits no approximate identity. Since $\langle f \rangle$ is a closed ideal in $L^1(G)$, under the $L^1$-norm, $\langle f \rangle$ forms a homogeneous Banach algebra which is a candidate. //

We will give a criterion on those homogeneous Banach algebras which admit an identity.

7.6. <u>THEOREM</u> Let $B(G)$ be a homogeneous Banach algebra. Then $\mu$ is an identity of $B(G)$ if and only if $\mu \in M(G)$ with $\hat{\mu} = \chi_\Delta$ where $\Delta$ is a compact set in $\Gamma$ belonging to the coset ring of $\Gamma$ such that $\Delta = Z(B(G))^c$.

<u>PROOF</u>. "Only-if-part" Suppose $\mu$ is an identity of $B(G)$. Then $\mu$ is an idempotent measure. By Theorem 1.13, $\hat{\mu} = \chi_\Delta$ where $\Delta$ belongs to the coset ring of $\Gamma$. The compactness of $\Delta$ follows from the Riemann-Lebesgue Lemma. Plainly, $\Delta = Z(B(G))^c$.

"If-part" Suppose $\mu \in M(G)$ with $\hat{\mu} = \chi_\Delta$ where $\Delta$ is a compact set in $\Gamma$ belonging to the coset ring of $\Gamma$ such that $\Delta = Z(B(G))^c$. Pick a function $f$ in $L^1(G)$ such that

$\hat{f}(\Delta) = 1$ . Then $\hat{\mu} = \hat{\mu}\hat{f}$ . Consequently $\mu = \mu * f \in L^1(G)$ . Moreover, $\mu$ is a function in $L^1(G)$ such that Supp $\hat{\mu}$ is compact and disjoint with $Z(B(G))$ , $\mu \in B(G)$ by Theorem 1.15. Clearly $\mu$ is an identity in $B(G)$ .

7.7. <u>THEOREM</u> Let $(B, \|\ \|)$ be a separable Banach algebra. Then $B$ admits a bounded left approximate identity if and only if $B$ admits a strong bounded left approximate identity.

<u>PROOF.</u> It suffices to show that $B$ admitting a bounded left approximate identity implies $B$ admitting a strong bounded left approximate identity. Let $\{x_k\}_{k=1}^{\infty}$ be a dense subset of $B$ . There is a constant $C > 0$ such that for each positive integer $n$ , there exists $e_n$ such that

$$\|e_n\| \le C$$

and

$$\|e_n x_k - x_k\| < \frac{1}{n} \quad \text{for} \quad k = 1, 2, \cdots, n .$$

Fixed $k$ , we have $\lim_{n\to\infty} \|e_n x_k - x_k\| = 0$ . Given $x \in B$ and $\varepsilon > 0$ , take $x_k$ such that

$$\| x_k - x \| < \frac{\varepsilon}{3(C+1)} .$$

Now

$$\begin{aligned} 0 &\le \overline{\lim_{n\to\infty}} \|e_n x - x\| \\ &\le \overline{\lim_{n\to\infty}} \|e_n x - e_n x_k\| + \overline{\lim_{n\to\infty}} \|e_n x_k - x_k\| + \|x_k - x\| \\ &< \frac{\varepsilon}{3} + \frac{\varepsilon}{3} + \frac{\varepsilon}{3} = \varepsilon . \end{aligned}$$

Since $\varepsilon$ is arbitrary, $\lim_{n\to\infty} \|e_n x - x\| = 0$ . This asserts that B admits a strong bounded left approximate identity. //

The preceding theorem can be applied to the function algebras. Let X be a locally compact Hausdorff space, and $C_0(X)$ the function algebra, which is a Banach algebra with pointwise multiplication, of continuous functions on X vanishing at infinity, under the sup norm $\| \ \|_\infty$ . If X is compact, we write C(X) instead of $C_0(X)$ .

7.8. <u>THEOREM</u>

(i) $C_0(X)$ is separable if and only if X is metrizable and $\sigma$-compact.

(ii) $C_0(X)$ admits a bounded approximate identity.

(iii) $C_0(X)$ admits a strong bounded approximate identity if and only if X is $\sigma$-compact.

(iv) There exists a non-separable Banach algebra admitting a strong bounded approximate identity.

<u>PROOF</u>.

(i) Let $X' = X \cup \{\infty\}$ be the one point compactification of X . Then each f in $C_0(X)$ can be identifies with a function $\tilde{f}$ in $C(X')$ such that $\tilde{f}|_X = f$ , and $\tilde{f}(\infty) = 0$ . We claim that $C_0(X)$ is separable if and only if so is $C(X')$ . In fact, if $C(X')$ is separable, then, as an open subspace of $C(X')$ , $C_0(X)$ is separable. Conversely, if $C_0(X)$ is separable, then take dense subset $\{f_n\}_{n=1}^{\infty}$

of $C_o(X)$ . For $r = a + ib$ where $a$ and $b$ are rationals, and $n = 1,2,\cdots$, define

$$g_{r,n} = r + \tilde{f}_n \ .$$

Obviously, the set of all $g_{r,n}$ is dense in $C(X')$ . Thus $C(X')$ is separable. Since $X'$ is compact, $C(X')$ is separable if and only if $X'$ is metrizable, and that $X'$ is metrizable if and only if $X$ is metrizable and $\sigma$-compact. Therefore $C_o(X)$ is separable if and only if $X$ is metrizable and $\sigma$-compact.

(ii) Let $f_1, f_2, \cdots, f_n$ be any $n$ functions in $C_o(X)$ , and $\varepsilon > 0$ . Take $n$ compact sets $K_1, K_2, \cdots, K_n$ in $X$ such that

$$\sup_{x \varepsilon K_i^c} |f_i(x)| < \frac{\varepsilon}{4} \quad \text{for} \quad i = 1,2,\cdots,n \ .$$

Let $f \varepsilon C_o(X)$ such that $f(K_1 \cup K_2 \cup \cdots \cup K_n) = 1$ , and $|f(x)| \leq 1$ for $x \varepsilon X$ . Then

$$||f||_\infty \leq 1$$

and

$$||ff_i - f_i||_\infty \leq \frac{\varepsilon}{2} < \varepsilon \quad \text{for} \quad i = 1,2,\cdots,n \ .$$

Therefore $C_o(X)$ admits a bounded approximate identity.

(iii) Suppose $X$ is $\sigma$-compact. Take a sequence $(K_n)_{n=1}^\infty$ of compact sets in $X$ such that $K_n$ is contained in the interior of $K_{n+1}$ for each $n$ and $X = \bigcup_{n=1}^{\infty} K_n$ . Let $f_n$ be in $C_o(X)$ such that $0 \leq f_n \leq 1$ , $f_n(K_n) = 1$, and $f_n(K^c_{n+1}) = 0$ , for each $n$ . Routine arguments reveal that $(f_n)$ is a strong bounded approximate identity in $C_o(X)$ .

Conversely, assume that $X$ is not a $\sigma$-compact but nevertheless $C_o(X)$ admits a strong bounded approximate identity $(f_n)_{n=1}^{\infty}$ , say. For each $n$ , take a compact set $K_n$ such that

$$\sup_{x \varepsilon K_n^c} |f_n(x)| < \frac{1}{n} .$$

Let $a \varepsilon X \setminus \bigcup_{n=1}^{\infty} K_n$ . This is possible since $X$ is not $\sigma$-compact. Let $K$ be compact neighborhood of $a$ , and $f \varepsilon C_o(X)$ such that $f(a) = 1$ and $f(K^c) = 0$ . Now

$$\begin{aligned} \|f_n f - f\|_\infty &\geq |f_n(a)f(a) - f(a)| \\ &= 1 - |f_n(a)| \\ &> 1 - \frac{1}{n} \quad \text{for } n = 1,2,\cdots . \end{aligned}$$

This contradicts that $\lim_{n\to\infty} \|f_n f - f\|_\infty = 0$ . Therefore if $C_o(X)$ admits a strong bounded approximate identity, then $X$ is $\sigma$-compact.

(iv) Let $X$ be a non-metrizable $\sigma$-compact space. Then $C_o(X)$ is a non-separable Banach algebra admitting a strong bounded approximate identity. //

An interesting result is as follows:

7.9. <u>THEOREM</u> Let $(B, \| \ \|)$ be a reflexive Banach algebra.

Then $B$ admits a left identity if and only if $B$ admits a bounded left approximate identity.

PROOF. It suffices to show that if $B$ admits a bounded left approximate identity $(e_\lambda)_{\lambda\varepsilon\Lambda}$, say, then $B$ admits a left identity. Since $B$ is reflexive, identified it as a subset of $B^{**}$, $\{e_\lambda\}_{\lambda\varepsilon\Lambda}$ is a bounded set in $B^{**}$. By the Alaoglu Theorem, there exists $e \varepsilon B$ and a subset $(e_{\lambda_\beta})$ of $(e_\lambda)$ such that

$$e_{\lambda_\beta} \to e \quad \text{in the weak-topology.}$$

Given $f \varepsilon B^*$, $x \varepsilon B$, define

$$g(y) = f(yx) \qquad (y \varepsilon B) .$$

Then $g \varepsilon B^*$ since

$$\begin{aligned} |g(y)| &= |f(yx)| \\ &\leq \|f\| \, \|x\| \, \|y\| . \end{aligned}$$

Or

$$g(e_{\lambda_\beta}) \to g(e) .$$

That is

$$f(e_{\lambda_\beta} x) \to f(ex) .$$

On the other hand, since $e_{\lambda_\beta} x \to x$,

$$f(e_{\lambda_\beta} x) \to f(x) .$$

Hence

$$f(x) = f(ex) .$$

This is valid for all $f \in B^*$, so $x = ex$. //

An application of the preceding theorem is to the operator algebras, which is a Banach algebra of bounded linear operators on a Hilbert space with composition as multiplication. Two special important operator algebras are B*-algebras and H*-algebras, where a B*-algebra admits a bounded approximate identity and an H*-algebra is also a Hilbert space. For these terminologies and their elementary properties, refer to Rickart (1960, §8, pp.272-276).

**7.10.** <u>COROLLARY</u>

(i) A reflexive B*-algebra admits a left identity.

(ii) An H*-algebra admits a left identity if it admits a bounded left approximate identity.

## 8. FACTORIZATION AND NONFACTORIZATION

Before we are goint to describe the factorization in Banach algebras, we see some definitions:

**8.1.** <u>DEFINITION</u> Let $B$ be a Banach algebra.

(1) $B$ is said to have <u>feeble factorization</u> if, given $f \in B$, there are $f_i, g_i \in B$, $i = 1,2,\cdots$ so that

$$f = \sum_{i=1}^{\infty} f_i g_i \quad \text{(converging in the B-norm)} .$$

(2) $B$ is said to have <u>weak factorization</u> if, given $f \in B$, there are $f_1,\cdots,f_n$, $g_1,\cdots,g_n \in B$ so that

$$f = \sum_{i=1}^{n} f_i g_i \ .$$

(3) B is said to have factorization if, given $f \in B$, there are $g,h \in B$, so that

$$f = gh \ .$$

(4) B is said to have Cohen factorization if, there exists a constant $d$, such that, for $f \in B$ and $\varepsilon > 0$ there are $g,h \in B$ so that

$$f = gh$$

$$\|f - h\| < \varepsilon$$

$$\|g\| \le d \ .$$

It is clear that

Cohen factorization $\Longrightarrow$ factorization

$\Longrightarrow$ weak factorization

$\Longrightarrow$ feeble factorization .

Moreover, Lakien (1975) proved.

**8.2.** THEOREM Let S be a Segal algebra on an infinite compact abelian group G, with weak factorization. Then there is a constant K so that given $f \in P(S)$ there are $f_i$, $g_i \in B$, $i = 1,2,\cdots$ so that $f = \sum_{i=1}^{\infty} f_i * g_i$ and $\sum_{i=1}^{\infty} \|f_i\| \, \|g_i\| \le K\|f\|$.

The factorization problem was initiated by Salem (1945), who proved

$$L^1(T) * L^1(T) = L^1(T) \ , \quad L^1(T) * C(T) = C(T) \ .$$

Rudin (1957,58) followed to prove

$$L^1(G) * L^1(G) = L^1(G) \quad \text{for} \quad G = R \text{ , or a locally compact Euclidean abelian group .}$$

Rudin's methods can not extend to the case of arbitrary groups because they depend upon the use of the Fourier transform and particular functions in Euclidean n-space. The decisive step in all factorization theorems was taken by Cohen (1950):

8.3. <u>THEOREM</u> [Cohen(1950)] A Banach algebra with a bounded left approximate identity admits the factorization property.

As a consequence of Cohen's Theorem, we have

$$L^1(G) * L^1(G) = L^1(G)$$

for any locally compact group $G$ .

Cohen's Theorem has important applications throughout other parts of harmonic analysis. Making use of Cohen's construction, Varopoulos (1964 [1], [2]) proved:

8.4. <u>THEOREM</u> If $A$ is a Banach *-algebra with a bounded left approximate identity, then every positive linear functional on $A$ is continuous. In particular, every positive linear functional on $L^1(G)$ is continuous.

The other application of Cohen's ideal is the Module Factorization Theorem, which were made by Hewitt (1964), and independently by Curtis and Figà-Talamanca (1966):

8.5. <u>THEOREM</u> [Module Factoization] Let $(A, \| \|_A)$ be a Banach algebra with a bounded left approximate identity and $(M, \| \|_M)$ a Banach A-module. If AM is dense in M, then

$$AM = M .$$

In particular,

$$L^1(G) * L^p(G) = L^p(G) , \qquad L^1(G) * C_o(G) = C_o(G) .$$

Cohen's factorization theorem initiated the study of factorization problems in various Banach algebras and Banach modules. First of all, Cohen's theorem has a conversion.

8.6. <u>THEOREM</u> [Altman (1975)] A Banach algebra has the Cohen factorization property if and only if it has a bounded (or bounded weak) left approximate identity.

<u>It is natural to ask if both the Cohen factorization property and the (algebraic) factorization property are equivalent</u>, or the factorization property is the necessary and sufficient conditions of the possession of a bounded left approximate identity. <u>Paschke (1973) answer this question negatively</u>.

8.7. <u>THEOREM</u> Let B be a Banach algebra with identity 1, and suppose $a \in B$ satisfies:

(i) $\| a \| = 1$ ,

(ii) for $b \in B$ , $ab = 0$ implies $b = 0$ ,

(iii) $aB$ is not closed in B ,

(iv) $b_1 a b_2 = 1$ for some $b_1, b_2 \in B$ .

Let $A = aB$ . Then A can be renormed to be a Banach algebra admitting the factorization property, but which admits neither bounded left nor bounded right approximate identity.

PROOF. Define $|| \ ||_0$ on $A$ by $||ab||_0 = ||b||$ for $b \in B$. This is a well-defined Banach space norm on $A$ by (ii). For $b,c \in B$, we have $||(ab)(ac)||_0 = ||bac|| \leq ||b|| \, ||c|| = ||ab||_0 ||ac||_0$, so $|| \ ||_0$ is an algebra norm on $A$ as well. The algebra $B$ is naturally a left A-module, and in fact a Banach A-module, since for $b,c \in B$, $||(ab)c|| \leq ||b|| \, ||c|| = ||ab||_0 ||c||$. By (iii), $A \cdot B = aB$ is not a closed subspace of $B$; we conclude from Module Factorization Theorem that $A$ does not admit bounded left approximate identity. Suppose that $A$ admits a right approximate identity (bounded or otherwise) $(ae_\lambda)$. Then in particular $\lim_\lambda ||ae_\lambda - 1|| = \lim_\lambda ||a^2 e_\lambda - a||_0 = 0$, showing that $\overline{aB} = B$ and hence $aB = B$, since a Banach algebra with identity can have no dense proper right or left ideals. This contradicts (iii), so $A$ doesn't admit right approximate identity. Finally, we note that (iv) allows us to factor any element $ab$ of $A$ as $ab = (abb_1)(ab_2)$. //

We now give an example of a situation in which the hypotheses of the lemma are satisfied. Let $\ell^2$ denote the space of all absolutely square-summable sequences of complex numbers, normed in the usual way, and let $B$ be the Banach algebra of all bounded linear operators on $\ell^2$. Let $(\lambda_n)$ be a sequence of real numbers with $0 < \lambda_n \leq 1$ for $n = 1,2,\cdots$, and

$\lim_n \lambda_n = 0$ . Letting $x_j$ denote the $j^{th}$ coordinate of the element $x \in \ell^2$ , we define operators $U, V \in B$ by

$$Ux = (x_1, \lambda_1 x_2, x_3, \lambda_2 x_4, x_5, \cdots)$$

$$Vx = (x_1, 0, x_2, 0, x_3, 0, \cdots\cdots) .$$

Clearly $||U|| = 1$ and $U$ is one-one, so for $W \in B$ , $UW = 0$ implies $W = 0$ . For $n = 1,2,\cdots$, let $E_n \in B$ be the orthogonal projection on the (2n)th coordinate. We have $\lim_n ||UE_n|| = \lim_n \lambda_n = 0$ . Since $||E_n|| = 1$ for each $n$, we see from the Open Mapping Theorem that $UB$ cannot be closed in $B$ . A direct computation shows that $V^*UV = 1$ . We may now invoke the theorem (with $V^*$ , $U$ , and $V$ playing the roles of $b_1$, $a$, and $b_2$ , respectively) to see that $UB$ , appropriately normed, is a Banach algebra which admits the factorization property but admits neither bounded left nor bounded right approximate identity.

The factorization property of $L^1(G)$ stimulated to investigate that property of various group algebras, especially to investigate the factorization property of Segal algebras.

There is a long and interesting history for the nonfactorization property of Segal algebras. Some Segal algebras obviously do no have the factorization such as $L^p(G)$ , $2 \leq p < \infty$ and $G$ an infinite compact abelian group, do not have the factorization property since

$$L^p(G) * L^q(G) \subset C(G) \subsetneqq L^p(G) \qquad (\frac{1}{p} + \frac{1}{q} = 1) .$$

The Segal algebras $f * S$ obviously does not have the factorization property since $S = f * S$ iff $f$ is the identity of $M(G)$, and $S \supsetneqq f * S \Rightarrow (f * S) * (f * S) \subseteq f * f * S \subsetneqq f * S$. Moreover, Edwards (1965) proved the Segal algebras $L^p(G)$, $1 < p < \infty$, do not have the <u>feeble factorization</u> property. The theorems for the nonfactorization of the noncompact case were first studied by Yap (1970) and Martin-Yap (1970). They proved the algebras $L^1 \cap L^p(G)$, $1 < p < \infty$, and $A^p(G)$, $1 \leq p < \infty$, do not have the factorization property, respectively, where $G$ is a non-discrete locally compact abelian group. Wang (1972) proved:

8.8. <u>THEOREM</u> [Wang (1972)] An $F^\mu P^\mu$-algebra doesn't admit the weak factorization property.

The main ideas of the proof are as follows:

(i) If $A(G) \subset L^1(G)$ admits the weak factorization property and if $\widehat{A(G)} \subset L^p(\mu)$ for some $p$, $0 < p < \infty$, and for a positive regular measure $\mu$ on $\Gamma$, then $\widehat{A(G)} \subset L^p(\mu)$ for all $p$, $0 < p < \infty$.

(ii) But if $A(G)$ is a $P^\mu$-algebra as well, then there is a $f \in A(G)$ such that $\hat{f} \notin L^r(\mu)$ for some $r$, $0 < r < \infty$.

(iii) Hence $A(G)$ does not admit the weak factorization property.

PROOF. I. If $A(G)$ is an $F^{\mu}$-algebra with the weak factorization, then $\widehat{A(G)} \subset L^{p}(\mu)$ for all $p$, $0 < p < \infty$.

Since $A(G)$ is an $F^{\mu}$-algebra, take $P_0$, $0 < P_0 < \infty$ such that $\widehat{A(G)} \subset L^{P_0}(\mu)$. But $\widehat{A(G)} \subset L^{\infty}(\mu)$, so $A(G) \subset L^{p}(\mu)$ for all $p$, $P_0 \leq p \leq \infty$. Given $f \in A(G)$, there exist $g_i$, $h_i \in A(G)$, $i = 1,2,\cdots,n$ such that $f = \sum_{i=1}^{n} g_i * h_i$. Consequently, $\hat{f} = \sum_{i=1}^{n} \hat{g}_i\hat{h}_i$. By the Cauchy inequality, for $i = 1, 2,\cdots,n$, we have

$$\int |\hat{g}_i\hat{h}_i|^{\frac{P_0}{2}} d\mu = \int |\hat{g}_i|^{\frac{P_0}{2}} |\hat{h}_i|^{\frac{P_0}{2}} d\mu$$
$$\leq \left(\int |\hat{g}_i|^{P_0} d\mu\right)^{\frac{1}{2}} \left(\int |\hat{h}_i|^{P_0} d\mu\right)^{\frac{1}{2}}$$
$$< \infty .$$

This asserts that $\hat{g}_i\hat{h}_i \in L^{\frac{P_0}{2}}(\mu)$ for $i = 1,2,\cdots,n$. It turns out that $f \in L^{\frac{P_0}{2}}(\mu)$. Hence $\widehat{A(G)} \subset L^{\frac{P_0}{2}}(\mu)$. Continuing this process, we have $\widehat{A(G)} \subset L^{\frac{P_0}{2^n}}(\mu)$ for $n = 1,2,\cdots$. But for arbitrary $P$, $0 < p < P_0$, there exists an integer $N$ such that $\frac{P_0}{2^N} \leq p < P_0$. $\widehat{A(G)} \subset L^{\frac{P_0}{2^N}}(\mu) \cap L^{p}(\mu)$ implies $\widehat{A(G)} \subset L^{p}(\mu)$. Our assertion then follows.

II. If $A(G)$ is a $P^{\mu}$-algebra, then there is $f \in A(G)$ such that $\hat{f} \notin L^{r}(\mu)$ for some $r$, $0 < r < \infty$. Notations are as in Definition 4.5. Since

$$\sum_{n=1}^{\infty} \frac{\|f_n\|_A}{C_n^{a+1}} \le \sum_{n=1}^{\infty} \frac{C_n}{C_n^{a+1}}$$

$$= \sum_{n=1}^{\infty} \frac{1}{C_n^a} < \infty$$

there exists $f \in A(G)$ with $f = \sum_{n=1}^{\infty} \frac{f_n}{C_n^{a+1}}$, the series converging with respect to $\| \ \|_A$. That $A(G)$ is subalgebra of $L^1(G)$ implies $\|g\|_{L^1} \le C\|g\|_A$ for some constant $C > 0$ and for all $g$ in $A(G)$. Consequently, $\hat{f} = \sum \frac{\hat{f}_n}{C_n^{a+1}}$. However, the supports of $\hat{f}_n$ are pairwise disjoint, so

$$|\hat{f}|^{\frac{b}{a+1}} = \sum \frac{|\hat{f}_n|^{\frac{b}{a+1}}}{C_n^b} .$$

It turns out

$$\int_\Gamma |\hat{f}|^{\frac{b}{a+1}} \, d\mu = \sum_{n=1}^{\infty} \int_\Gamma \frac{|\hat{f}_n|^{\frac{b}{a+1}}}{C_n^b} \, d\mu$$

$$\ge \sum_{n=1}^{\infty} \int_{\theta_n} \frac{|\hat{f}_n|^{\frac{b}{a+1}}}{C_n^b} \, d\mu$$

$$= \sum_{n=1}^{\infty} \int_{\theta_n} \frac{1}{C_n^b} \, d\mu$$

$$= \sum_{n=1}^{\infty} \frac{\mu(\theta_n)}{C_n^b} = \sum_{n=1}^{\infty} \frac{\alpha}{C_n^b}$$

$$= \infty .$$

Hence $\hat{f} \notin L^r(\mu)$ where $r = \frac{b}{a+1}$ , $0 < r < \infty$ .

By I and II, an $F^\mu P^\mu$-algebra does not admit the weak factorization property. //

8.9. <u>REMARK</u> From the preceding proof, it is easy to see that an $F^\mu$-algebra $A(G)$ , in which there is $f \in A(G)$ such that $\hat{f} \notin L^p(\mu)$ , for some $p$ , $0 < p < \infty$ , does not admit the weak factorization property.

8.10. <u>THEOREM</u> [Wang (1972)] Let $S(G)$ be a character Segal algebra such that $\widehat{S(G)} \subset L^p(\Gamma)$ for some $p$ , $0 < p < \infty$ . Then $S(G)$ does not admit the weak factorization property.

<u>PROOF.</u> Recall that a character Segal algebra is a P-algebra. //

8.11. <u>COROLLARY</u> Let $S(T)$ be a Segal algebra containing $C^\infty(T)$ such that $\widehat{S(T)} \subset L^p(\Gamma)$ for some $p$ . $0 < p < \infty$ . Then $S(G)$ does not admit the weak factorization property.

<u>PROOF.</u> Recall that by Corollary 4.11, a Segal algebra on $T$ containing $C^\infty(T)$ is a P-algebra. //

8.12. <u>COROLLARY</u> The algebras $C^k(T)$ , $L^{(k)}(R)$ , $C(G)$ , $L^p(G)$ , $W(G)$ , $L^1 \cap C_o(G)$ , $L^1 \cap L^p(G)$ , $A^p(G)$ , $Lip_\alpha(T)$ , and $BV(T)$ in sections 3 and 4 do not admit the weak factorization property.

<u>PROOF.</u> Recall that all of them are FP-algebras. //

8.13. <u>DEFINITIONS</u> Let $G$ be an infinite compact abelian group with character group $\Gamma$ . Then,

(i) A subset $\Delta$ of $\Gamma$ is a Sidon set if for every $f \in C(\Gamma)$ such that $f(\Delta^c) = 0$ , then there is $\mu \in M(G)$ such that $f = \hat{\mu}$ .

(ii) Let $\Delta$ be a subset of $\Gamma$ and $p$ a number, $1 \leq p < \infty$ . Then $\Delta$ is said to be a $\Lambda_p$ set if for every $f \in L^1(G)$ such that $\hat{f}(\Delta^c) = 0$ , then $f \in L^p(G)$ . This is equivalent to $L^1_\Delta(G) = L^p_\Delta(G)$ .

8.14. LEMMA [Hweitt-Ross (1970, p.423)] Let $\Delta$ be a Sidon set in $\Gamma$ . Then $\Delta$ is a $\Lambda_p$ set for all $p$ , $1 \leq p < \infty$ .

8.15. THEOREM Suppose that $G$ is an infinite compact abelian group, $\Delta$ an $\Lambda_p$-set and $1 < p < \infty$ . Then the homogeneous Banach algebra $L^1_\Delta(G)$ admits the weak factorization property if and only if $\Delta$ is a finite set.

PROOF. "If-part" . Suppose $\Delta$ is finite. Then $\Delta$ belongs to the coset ring (= D-coset ring) of $\Gamma$ . This asserts that $L^1_\Delta(G)$ admits a bounded approximate identity. In particular, $L^1_\Delta(G)$ admits the factorization property.

"Only-if-part" . Suppose that $\Delta$ is infinite. Since $L^1_\Delta(G) = L^p_\Delta(G)$ , $L^1_\Delta(G)$ is an F-algebra. Moreover, there is a function $f \in L^1_\Delta(G)$ such that $\hat{f} \notin L^{\frac{1}{2}}(\Gamma)$ . Let $(\gamma_k)_{k=1}^\infty$ be an infinite sequence in $\Delta$ . Since $\gamma_k \in L^1_\Delta(G)$ for $k=1,2,\cdots$ and

$$\sum_{k=1}^{\infty} \frac{\|\gamma_k\|_{L^1}}{k^2} = \sum_{k=1}^{\infty} \frac{1}{k^2} < \infty$$

there exists $f \varepsilon L^1_\Delta(G)$ such that

$$f = \sum_{k=1}^{\infty} \frac{\gamma_k}{k^2} .$$

Consequently,

$$\int_\Gamma |\hat{f}(\gamma)|^{\frac{1}{2}} d\gamma = \sum_{\gamma\varepsilon\Gamma} |\hat{f}(\gamma)|^{\frac{1}{2}}$$
$$= \sum_{k=1}^{\infty} \frac{1}{k}$$
$$= \infty.$$

Therefore $\hat{f} \not\varepsilon L^{\frac{1}{2}}(\Gamma)$ . By Remark 8.9, $L^1_\Delta(G)$ does not admit the weak factorization property. //

It is interesting to see that the homogeneous Banach algebra $L^1_\Omega(T)$ does not admit the weak factorization property, where $\Omega = \{0,1,2,\cdots\}$ . It is well-known that $L^1_\Omega(T)$ can be identified with the $H^1$-space, which is the space of all analytic functions $F$ in the open unit disc, such that

$$||F||_{H^1} = \sup_{0<\gamma<1} \frac{1}{2\pi} \int_0^{2\pi} |F(\gamma e^{it})| dt < \infty .$$

This identifiction shares the following property: Let $F \varepsilon H^1$, and write $F$ into the power series $F(z) = \sum a_n z^n$ , and let $f$ be the correspondent function in $L^1_\Omega(T)$ , then

$$\hat{f}(n) = a_n \quad \text{for} \quad n = 1,2,\cdots .$$

and

$$\sum_{n=1}^{\infty} \frac{|a_n|}{n} < \infty .$$

The last inequality was proved by Hardy [Katznelson (1968,p.91)]. Consider now the discrete regular measure $\mu$ on $Z$ defined by

$$\mu(n) = \frac{1}{n} \quad \text{for} \quad n = 1,2,\cdots .$$

Then $L^1_\Omega(T)$ is an $F^\mu$-algebra. That is, $\widehat{L^1_\Omega(T)} \subset L^1(\mu)$. We claim that $L^1_\Omega(T)$ does not admit the weak factorization property. For otherwise, $\widehat{L^1_\Omega(T)} \subset L^p(\mu)$ for <u>all</u> $p$, $0 < p \leq \infty$, along the same line as in the proof I of Theorem 8.8. However, pick up a function $F$ defined by

$$F(z) = \frac{1}{1-z}\left(\frac{1}{z}\log\frac{1}{1-z}\right)^{-2} .$$

Consequently $F \in H^1$, and if write

$$F(z) = \sum_{n=1}^{\infty} a_n z^n ,$$

and

$$a_n \sim \frac{C}{(\log n)^2} \quad \text{as} \quad n \to \infty$$

for some positive constant $C$ [Littlewood (1944,pp.93-96)]. We conclude that

$$\sum_{n=-\infty}^{\infty} |\hat{f}(n)|^{\frac{1}{2}} d\mu(n) = \sum_{n=1}^{\infty} \frac{|a_n|^{\frac{1}{2}}}{n}$$

$$\sim \sum_{n=1}^{\infty} \frac{1}{n \log n}$$

$$= \infty .$$

This asserts that $f \notin L^{\frac{1}{2}}(\mu)$. Consequently,

8.16. <u>THEOREM</u> The algebra $L^1_\Omega(T)$ does not admit the weak factorization property.

Varopoulos's Theorem and the nonfactorization give the following consideration: Suppose $B(G)$ is a group algebra; a Banach algebra of functions on $G$ with convolution as multiplication. For $f \in B(G)$, define

$$f^*(x) = \overline{f(-x)} .$$

If $f^* \in B(G)$ whenever $f \in B(G)$, then $B(G)$ is called stable under $*$, $f \to f^*$ is an involution, and under this "$*$" $B(G)$ forms a Banach $*$-algebra (for the definition see [Richart (1960, p.180)]. A linear functional $\sigma$ on $B(G)$ then is called positive if

$$\sigma(f^* * f) \geq 0 \quad \text{for} \quad f \in B(G) .$$

Varopoulos proved that every positive linear functional on $L^1(G)$ is continuous. We mention that the algebras in sections 3 and 4 such as $C^k(T)$, $L^{(k)}(T)$, $C(G)$, $L^p(G)$, $W(G)$, $L^1 \cap C_o(G)$, $L^1 \cap L^p(G)$, $A^p(G)$, $Lip_\alpha(T)$, $BV(T)$ are stable under $*$. A Segal algebra, which is not stable under $*$, was given in Theorem 4.19. Let $B(G)$ be any one of the above algebras which is stable under $*$. We know that $B(G)^2 \subsetneq B(G)$ while $B(G)^2$ is dense in $B(G)$. Then there exists a discontinuous positive linear functional on $B(G)$. As a matter of fact, we can prove a more general setting:

8.17. <u>THEOREM</u> Let $(A, \| \ \|_A)$ be a Banach $*$-algebra such that

$A^2 \subsetneqq A$ but $A^2$ is dense in $A$. Then there exists a discontinuous positive functional on $A$.

PROOF. The proof is quite elementary. Since $A^2$ is a linear subspace of $A$, by Zorn's Lemma there exists Hamel bases $E$ for $A^2$ and $D$ for $A$ such that $E \subsetneqq D$. Take $x_0 \in D \setminus E$ and let $M$ be the linear space spanned by $D \setminus \{x_0\}$. We conclude that $A = M \oplus \mathbb{C}\, x_0$, where $\mathbb{C}$ denotes the set of all complex numbers, and $M \supset A^2$. Given $x \in A$, there exists uniquely a $y \in M$ and $\alpha \in \mathbb{C}$ such that $x = y + \alpha x_0$, define $\sigma(x) = \alpha$. Then routine arguments reveal that $\sigma$ is a non-zero positive linear functional on $A$ such that $\sigma(A^2) \subset \sigma(M) = \{0\}$. Since $A^2$ is dense in $A$, $\sigma$ is discontinuous. //

Another nonfactorization theorem was discovered by Burnham (1972, [2]):

8.18. LEMMA Let $(X, \mathfrak{A}, \mu)$ be a positive measure space. Let $f$ be a bounded complex-valued function on $X$ with $f \in L^p(\mu)$ for some $p \in (1,\infty)$. If for each positive integer $n$, we can write $f = f_1 \cdots f_n$ (pointwise multiplication) with $f_i$ bounded and $f_i \in L^p(\mu)$, then $f \in L^1(\mu)$.

PROOF. By the hypothesis, $f \in L^t(\mu)$ for $t \geq p$. Choose

$t = n \geq p$ and $a_i = \frac{1}{n}$ . Since $|f_i|^t \in L^1$ we have by the generalized Holder inequality

$$\int |f| = \int (|f_1|^t)^{a_1} \cdots (|f_n|^t)^{a_n} \, d\mu$$

$$\leq \|f_1\|_1^{ta_1} \cdots \|f_n\|_1^{ta_1}$$

$$< \infty .$$

So $f \in L^1(\mu)$ .

8.19. <u>THEOREM</u> [Burnham] If $\hat{A} \not\subset L^1(\Gamma)$ but $\hat{A} \subset L^p(\Gamma)$ for some $p \in (1,\infty)$ , then $A$ does not have the factorization property.

<u>PROOF</u>. Suppose $A$ has the factorization property, then for each $g \in A$ and any positive integer $n$ we can write $g = g_1 * \cdots * g_n$ or $\hat{g} = \hat{g}_1 \cdots \hat{g}_n$ . By Lemma 8.18, $\hat{g} \in L^1(G)$ , a contradiction.

Leinert (1975) got a nonfactorization result which is analogous to Theorem 8.10.

8.20. <u>THEOREM</u> Suppose $\Gamma$ to be compactly generated and let $U = U^{-1}$ be a generating compact subset of $\Gamma$ . Let $B \subset L^1(G)$ be a subalgebra which is a Banach algebra with respect to some norm $\| \ \|_B$ and suppose that $\hat{B} \subset L^p(\Gamma)$

for some finite $p$ . Let $f \neq 0$ in $\hat{B}$ be such that $\hat{f}$ is nonnegative or has compact support in $\Gamma$ and such that all translates of $\hat{f}$ by elements of $\Gamma$ are in $\hat{B}$ . If there is a real polynomial $P$ such that

$$\|\chi_a \hat{f}\|_{\hat{B}} \leq P(n)$$

for all $a \in U^n$ and all $n \in N$ , then $B$ does not have the weak factorization property.

Lakien (1975) proved that <u>if $G$ is an infinite compact abelian group and $S(G)$ is a Segal algebra admitting the weak factorization, then its w-relative completion $\tilde{S}$ admits the weak factorization</u>. As a matter of fact, the converse holds:

8.21. <u>THEOREM</u> Let $S$ be an A-Segal algebra. If $\tilde{S}^A$ has the weak factorization, then $\tilde{S}^A = S$ . In particular, if $\tilde{S}^A$ has the weak factorization, then $S$ does so.

<u>PROOF</u>. Let $f \in \tilde{S}^A$ . Since $\tilde{S}^A$ has the weak factorization, then

$$f = \sum_{i=1}^{k} g_i * h_i$$

$$g_i,\ h_i \in \tilde{S}^A\ ,\ i = 1,2,\cdots,k\ .$$

Choose $(g_{in}) \subset S$ with

$$\|g_{in}\|_S \le M_i \quad \forall\alpha$$

$$g_{in} \xrightarrow{A} g_i \; .$$

Consider

$$\|\sum_{i=1}^{k} g_{in} * h_i - \sum_{i=1}^{k} g_i * h_i\|_{\tilde{S}}$$

$$\le \sum_{i=1}^{k} \|g_{in} - g_i\|_A \, \|h_i\|_{\tilde{S}}$$

$$\to 0 \quad \text{as} \quad n \to \infty \; .$$

Since $\sum_{i=1}^{k} g_{in} * h_i \in S$ and $S$ is closed in $\tilde{S}^A$, $f = \sum_{i=1}^{k} g_i * h_i \in S$. Thus $\tilde{S}^A = S$.

8.22. THEOREM Let $S(G)$ be a character Segal algebra and $\tilde{S}$ its w-relative completion. If $\tilde{S}$ has the weak factorization, then $S$ does do.

PROOF. Since $L^1 * \tilde{S} = S$,

$$\begin{aligned} S * S &= (L^1 * \tilde{S}) * (L^1 * \tilde{S}) \\ &= (L^1 * L^1) * (\tilde{S} * \tilde{S}) \\ &= L^1 * \tilde{S} \\ &= S \; . \end{aligned}$$

For the factorization, in addition to study group algebras, we also concern Banach-modules. Recall that Salem proved

$$L^1(T) * C(T) = C(T)$$

Later, Hewitt (1964) and, independently, Curtis-Figà-Talamanca (1966) proved a decisive Module Factorization Theorem:

8.23. THEOREM Let $(A, \|\ \|_A)$ be a Banach algebra admitting a left approximate identity bounded by $K$, and $(M, \|\ \|_M)$ a Banach A-module. If $AM$ is dense in $M$ then, for each $f \in M$ and $\varepsilon > 0$, there are $g \in A$, $h \in M$ with

$$f = g * h$$

$$\|g\|_A \leq K$$

$$\|f - h\|_M < \varepsilon .$$

For the proof see Hewitt-Ross (1970,p.268).

A consequence of Module Factorization Theroem is

$$L^1(G) * C_o(G) = C_o(G) , \quad L^1(G) * B(G) = B(G)$$

where $B$ is any homogeneous Banach algebra on $G$ (In particular, $B$ may be a Segal algebra).

Larsen (1973,[4]) established a Module Factorization Theroem:

8.24. THEOREM Let $S(G)$ be a Segal algebra on $G$. If $S * C_o(G) = C_o(G)$, then $\mathrm{Hom}_{L_1}(S,S) \simeq M(G)$.

$Hom_{L_1}(S,S)$ denotes the $L_1$-multiplier algebra of $S(G)$, and $\simeq$ the topological isomorphism.

As a consequence of the Theorem, we have

8.25. <u>COROLLARY</u>. (i) If $G$ is an infinite compact abelian group, then

(a) $A^p(G) * C(G) \subsetneqq C(G)$ , $1 \leq p < \infty$

(b) $L^p(G) * L_{p'}(G) \subsetneqq C(G)$ , $1 < p < \infty$ , $\frac{1}{p} + \frac{1}{p'} = 1$

(ii) $W(R) * C_o(R) \subsetneqq C_o(R)$

(iii) $L^{(1)}(R) * C_o(R) \subsetneqq C_o(R)$ .

Larsen (1974) also extablished an interesting theorem.

If $S$ is a Segal algebra and $W$ is an $L^1$-convolution module, then we define

$$S \underline{\otimes} W = \{w \in W : w = \sum_{k=1}^{\infty} g_k * w_k \ , \ (g_k) \subset S \ , \ (w_k) \subset W$$

$$\text{and} \ \sum_{k=1}^{\infty} \|g_k\|_S \|w_k\|_W < \infty\} .$$

$S \underline{\otimes} W$ is a Banach space with the norm

$$|||w||| = \inf \{ \sum_{i=1}^{\infty} \|g_k\|_S \|w_k\|_W \Big| w = \sum_{k=1}^{\infty} g_k * w_k \} .$$

Then we have

8.26. <u>THEOREM</u> If $L^1 * W = W$ , then the following are equivalent:

(i) $S \underline{\otimes} W = W$

(ii) $Hom_{L_1}(S,W^*) \simeq W^*$ .

8.27. <u>THEOREM</u> If $\mathrm{Hom}_{L_1}(S, M(G)) = \mathrm{Hom}_{L_1}(S,S)$ . Then the following are equivalent:

(i) $S \underline{\otimes} C_0(G) = C_0(G)$

(ii) $\mathrm{Hom}_{L_1}(S,S) \simeq M(G)$ .

The applications of above two theroems are

8.28. <u>THEOREM</u> Let $G$ be a nondiscrete locally compact abelian group.

(i) Let $1 \leq p < \infty$ , then

$A^p(G) \underline{\otimes} C_0(G) = C_0(G)$ if and only if $G$ is non-compact.

(ii) Let $1 < p < \infty$ , then $(L^1(G) \cap L^p(G)) \underline{\otimes} C_0(G) = C_0(G)$ if and only if $G$ is noncompact.

(iii) $(L^1(G) \cap C_0(G)) \underline{\otimes} C_0(G) = C_0(G)$ if and only if $G$ is noncompact.

After giving various positive and negative results on the factorization of Banach algebras, we mention here two open questions:

I. <u>Does a proper Segal algebra have the factorization property</u>?

Various partial solutions have been obtained: see Theorem 8.8, 8.10, 8.19, 8.20.

Recently, Leinert (1975) gave a proper A-Segal algebra (unnecessary commutative) which have the factorization. The construction is simple and nice:

Let $A = \ell^2(N)$ , $B = \ell^1(N)$ . Choose $\xi \varepsilon B$ with $||\xi||_A = 1$ and define

$$ab = \langle a,\xi\rangle b \qquad a,b \;\varepsilon\; A .$$

We obtain an associative multiplication for $A$ , with

$$||ab||_A \leq ||a||_A||b||_A \qquad \text{for} \quad a,b \;\varepsilon\; A .$$

Obviously we have $\xi b = b$ for all $b \;\varepsilon\; A$ .

For $a,b \;\varepsilon\; B$ we have

$$\begin{aligned} ||ab||_B &\leq ||a||_A||\xi||_A||b||_B \\ &\leq ||a||_B||b||_B \; . \end{aligned}$$

So $B$ is an abstract Segal algebra in $A$ having a left unit, and hence $B$ have the factorization property.

II. <u>Is there any proper Segal algebra $S(G)$ with</u>

$$\underline{S(G) * C_o(G) = C_o(G)} \; .$$

Refer to Larsen (1973, [4]) for informations.

# CHAPTER IV

# CLOSED SUBALGEBRAS OF HOMOGENEOUS BANACH ALGEBRAS

For compact abelian group $G$, all closed ideals in $L^1(G)$ are explicitly known: Each closed ideal is fully determined by knowledge of its zero set. This synthesis method first applied by Rudin (1962, Ch. 9) to study the closed subalgebras in $L^1(G)$. A natural conjecture is that a closed subalgebra $A$ of a homogeneous Banach algebra $B(G)$ is fully determined by its Rudin classes --- the equivalence classes induced by the equivalence relation $R$ where $\gamma_1 R \gamma_2$ if $\hat{f}(\gamma_1) = \hat{f}(\gamma_2)$ for all $f \in A$. However, this conjecture is shown to be false by an example due to Kahane (1965) when $B(G) = L^1(T)$ and due to Rider (1969) when $B(G) = L^p(T)$, $1 < p < 2$, while the conjecture is true due to Edwards (1967, II, p.16) when $B(G) = L^2(T)$.

In this chapter, we give a somewhat systematic study of closed subalgebras of homogeneous Banach algebras and Segal algebras. As a consequence, it is shown that, for the $A^p(T)$-algebra, that conjecture holds for $1 \leq p \leq 2$ but fails for $2 < p < \infty$; Theorem 9.6.

## 9. CLOSED SUBALGEBRAS OF HOMOGENEOUS BANACH ALGEBRAS

For the reader's convenience, we describe briefly the spectral synthesis theory of closed ideals in a homogeneous Banach algebra $B(G)$, where $G$ will denote an <u>infinite compact</u>

abelian group with character group $\Gamma$ . We say that the spectral synthesis holds for $B(G)$ if one of the following statements is satisfied:

(i) If $I$ and $J$ are two closed ideals in $B(G)$ with $Z(I) = Z(J)$ , then $I = J$ .

(ii) If $I$ is a closed ideal in $B(G)$ , and $f$ is a function in $B(G)$ such that $Z(I) \subset Z(f)$, then $f \in I$.

In the case (ii), if we denote $\langle I,f \rangle$ as the closed ideal generated by $I$ and $f$ , then obviously

$$Z(\langle I,f \rangle) = Z(I) .$$

The equivalence of (i) and (ii) then follows.

Recall that the algebra $L^1(G)$ satisfies one of the two equivalent conditions. In other words, the spectral synthesis holds for $L^1(G)$ . Moreover,

**9.1. THEOREM** The spectral synthesis holds for any homogeneous Banach algebra on compact abelian $G$ .

**PROOF.** Let $B(G)$ be a homogeneous Banach algebra and $I$ a closed ideal in $B(G)$ . We claim that if $\gamma \in \Gamma$ and $\gamma \notin Z(I)$, the $\gamma \in I$ . In fact, $\gamma \notin Z(I)$ implies that there is $g \in I$ such that $\hat{g}(\gamma) \neq 0$ . Recall that $g * \gamma = \hat{g}(\gamma)\gamma$ . Thus $\gamma = \frac{1}{\hat{g}(\gamma)} g * \gamma \in B(G)$ since $g \in B(G)$ , and $\gamma \in L^1(G)$ . It truns out $g * \gamma \in I$ , or $\gamma \in I$ . Suppose $f \in B(G)$ such that $Z(I) \subset Z(f)$ . By Theorem 3.7 (iv) , $B(G)$ admits an approximate identity with compactly supoorted Fourier transforms. For $\varepsilon > 0$, take $g \in P(B(G))$ such that

$$\|g * f - f\|_B < \varepsilon .$$

Let $g = \sum_{i=1}^{n} c_i \gamma_i$ , where $c_i$ are complex numbers and $\gamma_i \in \Gamma$. Since

$$g * f = \sum_{i=1}^{n} c_i \hat{f}(\gamma_i)\gamma_i ,$$

and that $\hat{f}(\gamma_i) \neq 0$ implies $\gamma_i \in I$ , we have $g * f \in I$ . We conclude that $f \in I$ . //

9.2. <u>REMARK</u> Let $B(G)$ be a homogeneous Banach algebra and I a closed ideal in $B(G)$ . The preceding proof tells us that if $\gamma \notin Z(I)$ , then $\gamma \in I$ . Consequently, suppose $\gamma_1 \neq \gamma_2$ with $\gamma_1 , \gamma_2 \notin Z(I)$ , then there is $f \in I$ such that $\hat{f}(\gamma_1) \neq \hat{f}(\gamma_2)$. In this situation, it needs only to take $f = \gamma_1$ or $\gamma_2$ . These lead to the following consideration: Define $\gamma_1 R \gamma_2$ if $\hat{f}(\gamma_1) = \hat{f}(\gamma_2)$ for all $f \in I$ . Then R is an equivalence relation in $\Gamma$ . One special equivalence class is $\Delta_0 = Z(I)$ which may be infinite, while the other equivalence classes denoted by $(\Delta_\alpha)$ are single. That the spectral synthesis holds for $B(G)$ is equivalent to that if for every closed ideal I , and for $f \in B(G)$ with $Z(I) \subset Z(f)$ , then f can be approximated in the $B(G)$-norm by the trigonometric polynomials P in I such that $\hat{P}$ is constant on each $\Delta_\alpha$ , $\alpha \neq 0$ .

Similarly, for any subalgebra (closed or not) A of a homogeneous Banach algebra $B(G)$ , write $\gamma_1 R \gamma_2$ if $\hat{f}(\gamma_1) = \hat{f}(\gamma_2)$ for $f \in A$ . The relation R is an equivalence relation in $\Gamma$ , induced by A . One distinguished equivalence class is $\Delta_0 = Z(A)$ ; $\Delta_0$ may be infinite. The other equivalence classes, denoted by $(\Delta_\lambda)$ , where $\lambda$ runs through a suitable index set $\Lambda$ , must be finite, since $\hat{f} \in C_0(\Gamma)$ for $f \in A$ and $\Gamma$ is discrete. We follow Kahane (1965) in terming each such $\Delta_\lambda$ ($\lambda = 0$ or not) a Rudin class of A . It is possible to carry the synthesis theory of closed ideals over the closed subalgebras. This assertion will follow from the following theorem of Rudin.

9.3. <u>THEOREM</u> Let B(G) be a homogeneous Banach algebra and A a closed subalgebra of B(G) . Suppose $(\Delta_\lambda)$ is the Rudin classes induced by A , and $P_\lambda$ , for each $\lambda \neq 0$ , is a trigonometric polynomial such that $\hat{P}_\lambda = \chi_{\Delta_\lambda}$ , the characteristic function of $\Delta_\lambda$ , then $P_\lambda \in A$ .

<u>PROOF</u>. We take from Edwards (1967, II, 11.3.2). For $\lambda \neq 0$ , take $f \in A$ such that $\hat{f}(\Delta_\lambda) = z_\lambda \neq 0$ . There are at most finite number of indices, $\lambda_1, \cdots \lambda_n$, say, distinct from $\lambda$ for which $\hat{f}(\Delta_{\lambda_i}) = z_\lambda$ . Let P be a trigonometric polynomial such that $\hat{P} = \chi_{\Delta_\lambda \cup \Delta_{\lambda_1} \cup \cdots \cup \Delta_{\lambda_n}}$ : By Theorem 1.15, $P \in B(G)$ . Moreover,

(i) $P \in A$

By the Riemann-Lebesgue Lemma, $\hat{f}(\Gamma)$ has no limit point other than zero. It follows that a polynomial F in one complex variable may be chosen so that

$$F(0) = 0 \ , \quad F(z_\lambda) = 1 \ , \quad |F(z)| < \frac{1}{2}$$

for $z \neq z_\lambda$ in $\hat{f}(\Gamma)$ . Indeed, we may choose a closed ball with center at zero and radius $\frac{|z_\lambda|}{2}$ which contains all elements in $\hat{f}(\Gamma)$ except at most finite number of elements $z_1, \cdots z_\ell$ , say. Put

$$F(z) = \left(\frac{z}{z_\lambda}\right)^N \frac{(z - z_1) \cdots (z - z_\ell)}{(z_\lambda - z_1) \cdots (z_\lambda - z_\ell)}$$

where the positive integer $N$ is to be chosen in a moment. Then, plainly, $F(0) = F(z_1) = \cdots = F(z_\ell) = 0$ , $F(z_\lambda) = 1$ , and if $z \neq z_\lambda$ in $\hat{f}(\Gamma)$ , then

$$|F(z)| = \left|\frac{z}{z_\lambda}\right|^N \frac{|z - z_1| \cdots |z - z_\ell|}{|(z_\lambda - z_1) \cdots (z_\lambda - z_\ell)|}$$

$$\leq \frac{(\frac{1}{2}|z_\lambda| + |z_1|) \cdots (\frac{1}{2}|z_\lambda| + |z_\ell|)}{2^N \ |(z_\lambda - z_1) \cdots (z_\lambda - z_\ell)|}$$

which can be made less than $\frac{1}{2}$ if $N$ is sufficiently large. Suppose that

$$F(z) = a_1 z + \cdots\cdots + a_N z^N ,$$

and that

$$g = a_1 f + \cdots\cdots + a_N f^N .$$

Then $g \in A$ , $\hat{g} = F \circ \hat{f}$ , and $\|\hat{g} - \hat{P}\|_\infty < \frac{1}{2}$ . Moreover, by the <u>spectral radius formula</u>.

$$\|\hat{g} - \hat{P}\|_\infty = \lim_{k\to\infty} \|(g - P)^k\|_B^{\frac{1}{k}} .$$

This implies

$$\|(g - P)^k\|_B < \frac{1}{2^k}$$

for sufficiently large $k$ . Obviously,

$$(\hat{g} - \hat{P})^k = \hat{g}^k - \hat{P}$$

Or

$$(g - P)^k = g^k - P .$$

We conclude

$$\|g^k - P\|_B < \frac{1}{2^k}$$

for sufficiently large $k$ . Since $A$ is a closed subalgebra of $B(G)$ and $g \varepsilon A$ , it appears that $P \varepsilon A$ .

(ii) Let $P_\lambda$ be a trigonometric polynomial such that $\hat{P}_\lambda = \chi_{\Delta_\lambda}$ . Take $h \varepsilon A$ such that $\hat{h}(\Delta_\lambda) \neq \hat{h}(\Delta_{\lambda_1})$ , and put

$$h_1 = \frac{h - \hat{h}(\Delta_{\lambda_1})P}{\hat{h}(\Delta_\lambda) - \hat{h}(\Delta_{\lambda_1})} .$$

By (i), $h_1 \varepsilon A$ , and plainly

$$\hat{h}_1(\Delta_\lambda) = 1 , \quad \hat{h}_1(\Delta_{\lambda_1}) = 0 .$$

Continuing this process, choose $h_i \varepsilon A$ such that

$$\hat{h}_i(\Delta_\lambda) = 1 , \quad \hat{h}_i(\Delta_{\lambda_i}) = 0$$

for $i = 1,2,\cdots,n$ . Clearly $P_\lambda = P * h_1 * \cdots * h_n \varepsilon A$ . This completes the proof. //

Suppose that $B(G)$ is a homogeneous Banach algebra and that $A$ is a subalgebra of $B(G)$ , which induces the Rudin classes $(\Delta_\lambda)$ . Let $P(A)$ denote the subalgebra of $B(G)$

generated by the trigonometric polynomials $P_\lambda$ such that $\hat{P}_\lambda = \chi_{\Delta_\lambda}$ , $\lambda \neq 0$ , (This notation coincides with the former one whenever $A = B(G)$) and let $A^{B(G)}$ denote the closed subalgebra of all $f$ in $B(G)$ such that $\hat{f}(\Delta_o) = 0$ , and $f$ is constant on each $\Delta_\lambda$ , $\lambda \neq 0$ . Note that $A^{B(G)} = A^{L^1}(G) \cap B(G)$, where $A^{L^1}(G) = \{f \in L^1(G) : \hat{f}(\Delta_o) = 0 , \hat{f}(\Delta_\lambda) = \text{constants} \quad \lambda \neq 0\}$. Then, plainly, $\overline{P(A)}^B$ and $A^{B(G)}$ are the minimal and the maximal closed subalgebras induces $(\Delta_\lambda)$ . More pricisely, $\overline{P(A)}^B$ and $A^{B(G)}$ induce the same Rudin classes $(\Delta_\lambda)$ and if $A_1$ is any closed subalgebra of $B(G)$ which induces $(\Delta_\lambda)$ , then $\overline{P(A)}^B \subset A_1 \subset A^{B(G)}$ . Furthermore, both $A$ and $\bar{A}^B$ evidently induce the same Rudin classes and therefore then $P(A) = P(\bar{A}^B)$ and $A^{B(G)} = (\bar{A}^B)^{B(G)}$ .

9.4. <u>DEFINITIONS</u> Let $B(G)$ be a homogeneous Banach algebra and $A$ a closed subalgebra of $B(G)$ , $A$ is called a R-subalgebra if one of the following equivalent conditions holds:

(i) $\overline{P(A)}^B = A = A^{B(G)}$ .

(ii) Suppose $f \in B(G)$ such that $\hat{f}(\Delta_o) = 0$ , and $\hat{f}$ is constant on each $\Delta_\lambda$ , $\lambda \neq 0$ , then $f$ can be approximatated in the $B(G)$-norm by trigonometric polynomials $P$ such that $\hat{P}$ is constant on each $\Delta_\lambda$ , $\lambda \neq 0$ ($(\Delta_\lambda)$ is the Rudin classes induced by A).

(iii) If $f \in B(G)$ such that $\hat{f}(\Delta_o) = 0$ and $\hat{f}$ is constant on each $\Delta_\lambda$ , $\lambda \neq 0$ , then $f \in \overline{P(A)}^B$ .

The R-synthesis holds for $B(G)$ if every closed subalgebra

of B(G) is a R-subalgebra. Otherwise, we say that the R-synthesis fails for B(G) .

Kahane (1965) proved that the R-synthesis fails for $L^1(T)$ and Rider (1969) proved that the R-synthesis fails for $L^p(T)$ , $1 < p < 2$ . As a matter of fact, Rider proved:

9.5. THEOREM There is a closed subalgebra A of $L^1(T)$ and a function f in $L^1(T)$ such that

(i) $f \in \bigcap_{1 \le p < 2} L^p(T)$

(ii) $f \in A^{L^1(T)}$ but $f \notin \overline{P(A)}^{L^1}$ .

The R-synthesis for many other homogeneous Banach algebras will be investigated:

9.6. THEOREM [Tseng-Wang (1975)] The R-synthesis holds for $A^p(T)$ , $1 \le p \le 2$ , but fails for $A^p(T)$ , $2 < p < \infty$ .

PROOF. (i) $1 \le p \le 2$ .

Let D be a closed subalgebra of $A^p(T)$ which induces the Rudin classes $(\Delta_\lambda)$ , and $f \in A^p(T)$ such that $\hat{f}(\Delta_0) = 0$, $\hat{f}(\Delta_\lambda) = z_\lambda \neq 0$ for $\lambda \neq 0$ . Suppose that the number of $\Delta_\lambda$ is $n_\lambda$ , $\lambda \neq 0$ . We conclude that

$$\sum_{\lambda \neq 0} n_\lambda |z_\lambda|^p = \sum_{n \in Z} |\hat{f}(n)|^p < \infty .$$

Consequently, there exists, for $\varepsilon > 0$ , a finite set $F_0$ of indices such that $(\sum_{\beta \in F^c} n_\beta |z_\beta|^p) < (\frac{\varepsilon}{2})^p$ whenever F is a finite set of indices with $F \supset F_0$ .

Let $P_\beta$ , $\beta \neq 0$ , be the trigonometric polynomial such that $\hat{P}_\beta = \chi_{\Delta_\beta}$ , and let $g = f - \sum_{\beta\varepsilon F} z_\beta P_\beta$ . Then, since $1 \leq p \leq 2$ , by the Hausdorff-Young Theorem,

$$\begin{aligned} \|g\|_{L^1} &\leq \|g\|_{L^q} \qquad (\tfrac{1}{p} + \tfrac{1}{q} = 1) \\ &\leq \|\hat{g}\|_{\ell^p} . \end{aligned}$$

But

$$\|g\|_{A^p} = \|g\|_{L^1} + \|\hat{g}\|_{\ell^p} .$$

So

$$\begin{aligned} \|g\|_{A^p} &\leq 2 \|\hat{g}\|_{\ell^p} \\ &= 2\Big( \sum_{n\varepsilon Z} |\hat{f}(n) - \sum_{\beta\varepsilon F} z_\beta \hat{P}_\beta(n)|^p\Big)^{\frac{1}{p}} \\ &= 2\Big( \sum_{\beta\varepsilon F^c} n_\beta |z_\beta|^p\Big)^{\frac{1}{p}} \\ &< \varepsilon . \end{aligned}$$

Or,

$$\|f - \sum_{\beta\varepsilon F} z_\beta P_\beta\|_{A^p} < \varepsilon .$$

Since $\varepsilon$ is arbitrary and $\sum_{\beta\varepsilon F} z_\beta P_\beta \ \varepsilon\ \overline{P(D)}^{A^p}$ , $f \ \varepsilon\ \overline{P(D)}^{A^p}$ . The R-synthesis therefore holds for $A^p(T)$ , $1 \leq p \leq 2$ .

(ii) $2 < p < \infty$.

Let $q$ be with $\frac{1}{p} + \frac{1}{q} = 1$ . Then $1 < q < 2$ . Hence, by the Hausdorff-Young Theorem, $L^q(T) \subset A^p(T)$ . By Theorem 9.5, there is a closed subalgebra $E$ of $L^1(T)$ and $f \ \varepsilon \bigcap_{1\leq q<2} L^q(T)$ such that $f \ \varepsilon\ E^{L^1}$ but $f \not\in \overline{P(E)}^{L^1}$ . Let $E$ induce the Rudin classes $(\Delta_\lambda)$ . Suppose that $2 < p < \infty$ .

$$D_p = \{g \in A^p(T) : \hat{g}(\Delta_o) = 0 \ , \ \hat{g}(\Delta_\lambda) = \text{constant for } \lambda \neq 0\} \ ,$$

then $D_p$ and $\overline{P(D)}^{A^p}$ are the maximal and minimal closed subalgebra of $A^p(T)$ inducing $(\Delta_\lambda)$ . We conclude that $P(D) = P(E)$ since $A^p(T)$ is a Segal algebra hence contains $P(E)$ . By the hypothesis, $f \in \bigcap_{1 \leq q < 2} L^q(T)$ . But, by the Hausdorff-Young Theorem, $L^q(T) \subset A^p(T)$ . Consequently, $f \in \bigcap_{2<p<\infty} A^p(T)$ , or $f \in D_p$ for $2 < p < \infty$ . But, nevertheless, $f \notin \overline{P(D)}^{A^p}$ since $\overline{P(D)}^{A^p} \subset \overline{P(D)}^{L^1} = \overline{P(E)}^{L^1}$ and $f \notin \overline{P(E)}^{L^1}$ . This proves part (ii) . //

Along the same line of the proof of Theorem 9.6 (i), we get the following more general result.

9.7. <u>THEOREM</u> The R-synthesis holds for $A^p(G)$ , $1 \leq p \leq 2$ .

We try to derive some conditions under which the spectral synthesis holds for a homogeneous Banach algebra.

9.8. <u>THEOREM</u> Let $B_1(G)$ and $B_2(G)$ be two homogeneous Banach algebras with $B_1(G) \subset B_2(G)$ and $A$ a R-subalgebra of $B_1(G)$ . Then $A = \bar{A}^{B_2} \cap B_1(G)$ . In particular, $I = \bar{I}^{L^1} \cap B_1(G)$ for any closed ideal $I$ in $B_1(G)$ .

<u>PROOF.</u> Since there is a constant $C > 0$ such that $\| \ \|_{B_2} \leq C \| \ \|_{B_1}$ , $J \cap B_1(G)$ is a closed subalgebra of $B_1(G)$ whenever $J$ is a closed subalgebra of $B_2(G)$ . Particularly,

$A^{B_2} \cap B_1(G)$ is a closed subalgebra of $B_1(G)$ . Furthermore, plainly,

$$\hat{f}(\gamma) = \hat{f}(\beta) \quad \text{for} \quad f \in \bar{A}^{B_2} \cap B_1(G)$$

$$\Longleftrightarrow \hat{g}(\gamma) = \hat{g}(\beta) \quad \text{for} \quad g \in A \ .$$

We conclude that both $\bar{A}^{B_2} \cap B_1(G)$ and $A$ induce the same Rudin classes. Thus $A = \bar{A}^{B_2} \cap B_1(G)$ .

Finally, for any closed ideal $I$ in $B_1(G)$ , $I = \bar{I}^{L^1} \cap B(G)$, since by Theorem 9.1, $I$ is a R-subalgebra. //

9.9. <u>THEOREM</u> Let $B_1(G)$ and $B_2(G)$ be two homogeneous Banach algebras with $B_1(G) \subset B_2(G)$ . Suppose that the R-synthesis holds for $B_2(G)$ , then the R-synthesis holds for $B_1(G)$ if and only if $A = \bar{A}^{B_2} \cap B_1(G)$ (or $A = \bar{A}^{L^1} \cap B_1(G)$) for any closed subalgebra $A$ of $B_1(G)$ .

<u>PROOF.</u> The only-if-part follows from Theorem 9.8. Suppose $A = \bar{A}^{B_2} \cap B_1(G)$ for any closed subalgebra $A$ of $B_1(G)$ . Let $A$ be a closed subalgebra of $B_1(G)$ inducing the Rudin classes $(\Delta_\lambda)$ . Then, plainly, we have the following equalities:

$$A^{B_1} = \{f \in B_1(G) : \hat{f}(\Delta_0) = 0,\ \hat{f}(\Delta_\lambda) = \text{constant},\ \lambda \neq 0\}$$

$$= \{f \in B_2(G) : \hat{f}(\Delta_0) = 0,\ \hat{f}(\Delta_\lambda) = \text{constant},\ \lambda \neq 0\} \cap B_1(G)$$

$$= \overline{P(A)}^{B_2} \cap B_1(G) \quad \text{since the R-synthesis holds for } B_2(G)$$

$$= \overline{\overline{P(A)}^{B_1}}^{B_2} \cap B_1(G) \left( \subset \overline{\overline{\overline{P(A)}^{B_1}}^{B_2}}^{L^1} \cap B_1(G) \subset \overline{\overline{P(A)}^{B_1}}^{L^1} \cap B_1(G) \right)$$

$$= \overline{P(A)}^{B_1} \quad \text{by the hypothesis .}$$

This asserts that A is a R-subalgebra and the theorem then follows. //

The R-synthesis for special homogeneous Banach algebras will be investigated.

**9.10. THEOREM** If A is not an R-subalgebra of $L^1(T)$, which induces the Rudin classes $(\Delta_\lambda)$, then the R-synthesis fails for the homogeneous Banach algebra $L^1_{\Delta_o^c}$.

**PROOF.** This assertion follows from that $\overline{P(A)}^{L^1}$, and $A^{L^1(T)}$ are closed subalgebras of $L^1_{\Delta_o^c}$, which induces the same Rudin classes but $\overline{P(A)}^{L^1} \subsetneqq A^{L^1(T)}$. //

It is interesting to note that as a homogeneous Banach algebra, $L^1_{\Delta_o^c}$ may lack of the R-synthesis property, while, as a subalgebra, $L^1_{\Delta_o^c}$ is a R-subalgebra of $L^1(T)$ since it is a closed ideal in $L^1(T)$; Theorem 9.1.

**9.11. THEOREM** Let $\Delta$ be a $\Lambda_p$-set in $\Gamma$, $1 \leq p < \infty$. Then the R-synthesis holds for the homogeneous Banach algebra $L^1_{\Delta^c}$.

**PROOF.** Since $L^2(T) = A^2(T)$, the R-synthesis holds for $L^2(T)$ by Theorem 9.6. We begin with the remark that if $\Delta$ is a $\Lambda_p$-set, then $L^1_{\Delta^c} \subset L^1 \cap L^2(T)$, and there is a constant C such that

$$\|f\|_{L^2} \leq C\|f\|_{L^1} \qquad (1)$$

for $f \in L^1_{\Delta^c}$ [see Edwards(1967, II, p.222)] Suppose that A is a closed subalgebra of $L^1_{\Delta^c}$, then $\overline{P(A)}^{L^1}$, A, and $A^{L^1(T)}$ induce the same Rudin classes. But, with (1), $\overline{P(A)}^{L^1} = \overline{P(A)}^{L^2}$ and $A^{L^1(T)} = A^{L^2(T)}$. We conclude that $\overline{P(A)}^{L^1}$ and $A^{L^1(T)}$ are two closed subalgebras of $L^2(T)$ which induce the same Rudin classes. Hence $\overline{P(A)}^{L^1} = A^{L^1(T)}$, or A is a R-subalgebra. //

## 10. CLOSED SUBALGEBRAS OF SEGAL ALGEBRAS

In this section, we'll devote to characterize some Segal algebras in which the R-synthesis holds, begining with a couple of interesting theorems.

10.1. <u>THEOREM</u> Let $S_1(G)$ and $S_2(G)$ be two Segal algebras with $S_1(G) \subset S_2(G)$. Suppose that the R-synthesis holds for both $S_1(G)$ and $S_2(G)$. Then $A \to \bar{A}^{S_2}$ is a 1-1 correspondence between the family of all closed subalgebras of $S_1(G)$ and that of $S_2(G)$. More precisely,

(i) For any two closed subalgebras $J_1$ and $J_2$ of $S_2(G)$, if $J_1 \cap S_1 = J_2 \cap S_1$, then $J_1 = J_2$;

(ii) For any two closed subalgebras $A_1$ and $A_2$ of $S_1(G)$ if $\bar{A}_1^{S_2} = \bar{A}_2^{S_2}$ then $A_1 = A_2$;

(iii) For any closed subalgebra J of $S_2(G)$, we have $J = \overline{J \cap S_1(G)}^{S_2}$.

<u>PROOF</u>. (i) Let $J_1$ and $J_2$ be two closed subalgebras of

$S_2(G)$ with $J_1 \cap S_1 = J_2 \cap S_1$. Then, by Theorem 4.2 (i), $P(J_1) \subset S_1(G)$. Consequently,

$$P(J_1) \subset J_1 \cap S_1 .$$

Or,

$$\begin{aligned} J_1 &= \overline{P(J_1)}^{S_2} \subset \overline{J_1 \cap S_1}^{S_2} \\ &= \overline{J_2 \cap S_1}^{S_2} \subset \overline{J_2}^{S_2} \\ &= J_2 . \end{aligned}$$

Similarly it can be shown that $J_2 \subset J_1$. Hence $J_1 = J_2$.

(ii) It suffices to show that $A = \bar{A}^{S_2} \cap S_1(G)$, for any closed subalgebra $A$ of $S_1(G)$. However, this assertion follows from Theorem 9.9.

(iii) Suppose that $J$ is a closed subalgebra of $S_2(G)$. Then $J \cap S_1(G)$ is a closed subalgebra of $S_1(G)$. By part (ii) ,

$$J \cap S_1(G) = \overline{J \cap S_1(G)}^{S_2} \cap S_1(G) .$$

Applied (i) to get

$$J = \overline{J \cap S_1(G)}^{S_2} .$$

Combining (i), (ii) and (iii), we have $A \to \bar{A}^{S_2}$ is a 1-1 map of the family of all closed subalgebras of $S_1(G)$ onto that of $S_2(G)$. //

**10.2.** <u>THEOREM</u> Let $S_1(G)$ and $S_2(G)$ be two Segal algebras with $S_1(G) \subset S_2(G)$. Suppose that the R-synthesis holds for

$S_1(G)$ . Then the R-synthesis holds for $S_2(G)$ if and only if $J = \overline{J \cap S_1(G)}^{S_2}$ for any closed subalgebra $J$ of $S_2(G)$.

PROOF. Suppose that the R-synthesis holds for $S_2(G)$ , then, by Theorem 10.1 (iii), $J = \overline{J \cap S_1(G)}^{S_2}$ for any closed subalgebra $J$ of $S_2(G)$ . On the other hand, suppose $J = \overline{J \cap S_1(G)}^{S_2}$ for any closed subalgebra $J$ of $S_2(G)$ . Then we claim that $P(J) = P(J \cap S_1(G))$ . In fact, for $\gamma, \beta \in \Gamma$ with $\hat{f}(\gamma) = \hat{f}(\beta)$ for $f \in J$ , then, plainly, $\hat{g}(\gamma) = \hat{g}(\beta)$ for $g \in J \cap S_1(G)$ . Conversely, let $\gamma, \beta \in \Gamma$ with $\hat{g}(\gamma) = \hat{g}(\beta)$ for $g \in J \cap S_1(G)$ . Suppose $f \in J$ , since $J = \overline{J \cap S_1(G)}^{S_2}$ , there is a sequence $(g_n)$ in $J \cap S_1(G)$ with $g_n \to f$ in the $S_2$-norm. It turns out $\hat{g}_n \to \hat{f}$ uniformly. Consequently, $\hat{f}(\gamma) = \hat{f}(\beta)$ for $f \in J$ . Now, since the R-synthesis holds for $S_1(G)$ ,

$$J \cap S_1(G) = \overline{P(J \cap S_1(G))}^{S_1} .$$

Therefore

$$\begin{aligned} J &= \overline{J \cap S_1(G)}^{S_2} \\ &= \overline{\overline{P(J \cap S_1(G))}^{S_1}}^{S_2} \\ &= \overline{P(J \cap S_1(G))}^{S_2} \\ &= \overline{P(J)}^{S_2} . \end{aligned}$$

Thus the R-synthesis holds for $S_2(G)$ . //

10.3. DEFINITION Two Segal algebras $(S_1(G), \|\ \|_{S_1})$ and $(S_2(G), \|\ \|_{S_2})$ are said to be comparable if there exists a constant $C > 0$ such that $\| \ \|_{S_1} \le C\| \ \|_{S_2}$ or $\| \ \|_{S_2} \le C\| \ \|_{S_1}$.

Note that two Segal algebras with one containing the other are comparable.

10.4. THEOREM Suppose $S_1(G)$ and $S_2(G)$ are two comparable Segal algebras. If the R-synthesis holds for both of them, then the R-synthesis holds for the Segal algebra $S_1 \cap S_2(G)$.

PROOF. Take a constant $C > 0$ such that $\| \ \|_{S_1} \le C\| \ \|_{S_2}$, say. Let $A$ be a closed subalgebra of $S_1 \cap S_2(G)$, and $f \in \bar{A}^{S_2} \cap (S_1 \cap S_2(G))$, pick up a sequence $(f_n)$ in $A$ such that $\|f_n - f\|_{S_2} \to 0$ as $n \to \infty$. Then, plainly, $\|f_n - f\|_{S_1} \to 0$ as $n \to \infty$. Consequently,

$$\|f_n - f\|_{S_1 \cap S_2} \to 0 \quad \text{as} \quad n \to \infty.$$

Or, $f \in \bar{A}^{S_1 \cap S_2} = A$. This asserts that $A = \bar{A}^{S_2} \cap (S_1 \cap S_2(G))$ for any closed subalgebra $A$ of $S_1 \cap S_2(G)$. The proof then follows from Theorem 9.9. //

By Theorems 10.1 and 10.4, we have

10.5. COROLLARY If $S_1(G)$ and $S_2(G)$ are two comparable Segal algebras in which the R-synthesis holds, then there is a 1-1 correspondence of the family of all closed subalgebras of $S_1(G)$ to that of $S_2(G)$.

We wish to recognize that the theories of closed subalgebras of two Segal algebras are the same if there is a 1-1 correspondence between the family of all closed subalgebras of one and that of the other.

10.6. <u>COROLLARY</u> The theories of closed subalgebras of the algebras $A^p(T)$, $1 \leq p \leq 2$, are the same.

In the final of this section, we shall study the algebras whose Rudin classes have bounded lengthes.

10.7. <u>THEOREM</u> [Kahane (1965)] Let $A$ be a closed subalgebra of $L^1(T)$ inducing the Rudin classes $(\Delta_\lambda)$. If there is a constant $C > 0$ such that $|n_1 - n_2| < C$ for all $n_1, n_2 \in \Delta_\lambda$, for all $\lambda \neq 0$. Then $A$ is an R-subalgebra.

As a matter of fact, we prove a more general setting.

10.8. <u>THEOREM</u> Let $S(T)$ be a Segal algebra containing $C(T)$. Suppose $A$ is a closed subalgebra of $S(T)$ which induces the Rudin classes $(\Delta_\lambda)$. If there is a constant $C > 0$ such that $|n_1 - n_2| < C$ whenever $n_1, n_2 \in \Delta_\lambda$ for all $\lambda \neq 0$, then $A$ is an R-subalgebra.

<u>PROOF.</u> Notes that $C(T) \subset S(T)$ implies there is a constant $K > 0$ with $\| \ \|_S \leq K \| \ \|_\infty$. Take the Féjer Kernal $(K_n)_{n=1}^{\infty}$ on $T$ [see Katznelson (1968, p.12)], and consider a function $f$ in $A^{S(T)}$. Let $f(\Delta_\lambda) = z_\lambda$ and $m_\lambda = \min_{j \in \Delta_\lambda} |j|$ for $\lambda \neq 0$. Then

$$K_n * f = \sum_{j=-n}^{n} \hat{K}_n(j)\hat{f}(j)e^{ijt}$$

$$= \sum_{\substack{\lambda \neq 0 \\ m_\lambda \leq n}} \sum_{j \varepsilon \Delta_\lambda} \hat{K}_n(j)\hat{f}(j)e^{ijt}$$

$$= \sum_{\substack{\lambda \neq 0 \\ m_\lambda \leq n}} z_\lambda \sum_{j \varepsilon \Delta_\lambda} \hat{K}_n(j)e^{ijt} .$$

Obviously

$$\sum_{\substack{\lambda \neq 0 \\ m_\lambda \leq n}} z_\lambda \sum_{j \varepsilon \Delta_\lambda} \hat{K}_n(m_\lambda)e^{ijt}$$

is a function in $P(A)$ . We have

$$||K_n * f - \sum_{\substack{\lambda \neq 0 \\ m_\lambda \leq n}} z_\lambda \sum_{j \varepsilon \Delta_\lambda} \hat{K}_n(m_\lambda)e^{ijt}||_S$$

$$= || \sum_{\substack{\lambda \neq 0 \\ m_\lambda \leq n}} z_\lambda \sum_{j \varepsilon \Delta_\lambda} [\hat{K}_n(j) - \hat{K}_n(m_\lambda)]e^{ijt}||_S$$

$$= || \sum_{\substack{\lambda \neq 0 \\ m_\lambda \leq n}} z_\lambda \sum_{j \varepsilon \Delta_\lambda} \frac{(m_j - |j|)}{n + 1} e^{ijt}||_S$$

$$\leq K \sum_{\substack{\lambda \neq 0 \\ m_\lambda \leq n}} |z_\lambda| \sum_{j \varepsilon \Delta_\lambda} \frac{|m_j - |j||}{n + 1}$$

since $||e^{ijt}||_S \leq K ||e^{ijt}||_\infty = K$

$$\leq C^2K \sum_{\substack{\lambda \neq 0 \\ m_\lambda \leq n}} \frac{|z_\lambda|}{n + 1}$$

$$\leq C^2K \sum_{j=-n}^{n} \frac{|\hat{f}(j)|}{n + 1} .$$

For $\varepsilon > 0$ there exists a positive integer $N_1$ such that

$$|j| \geq N_1 \qquad |\hat{f}(j)| \leq \frac{\varepsilon}{8C^2K} .$$

Then take a positive integer $N_2$ such that

$$\sum_{j=-N}^{N_1} \frac{|\hat{f}(j)|}{N_2 + 1} < \frac{\varepsilon}{8C^2K} .$$

For $n \geq N_1 + N_2$, we get

$$\begin{aligned} C^2K \sum_{j=-n}^{n} \frac{|\hat{f}(z)|}{n+1} &= C^2K \sum_{|j| \leq N_1} \frac{|\hat{f}(j)|}{n+1} + C^2K \sum_{|j| \geq N_1} \frac{|\hat{f}(j)|}{n+1} \\ &\leq C^2K \sum_{|j| \leq N_1} \frac{|\hat{f}(j)|}{N_2+1} + \frac{\varepsilon}{8} \sum_{|j| \geq N_1} \frac{1}{n+1} \\ &< \frac{\varepsilon}{8} + \frac{\varepsilon}{4} \\ &< \frac{\varepsilon}{2} . \end{aligned}$$

We conclude, for $\varepsilon > 0$ there is $N_1 + N_2 > 0$ such that

$$\|K_n * f - \sum_{\substack{\lambda \neq 0 \\ m_\lambda \leq n}} z_\lambda \sum_{j \varepsilon \Delta_\lambda} \hat{K}_n(m_\lambda) e^{ijt}\|_S < \varepsilon/2$$

whenever $n \geq N_1 + N_2$.

Recall that $K_n * f \to f$ in the S-norm. There is a positive integer $N_3$ such that, for $n \geq N_3$, we have

$$\|K_n * f - f\|_S < \frac{\varepsilon}{2} .$$

Consequently, pick up $n = N_1 + N_2 + N_3$, we have

$$\| f - \sum_{\substack{\lambda \neq 0 \\ m_\lambda \leq n}} z_\lambda \sum_{j \varepsilon \Delta_\lambda} \hat{K}_n(m_\lambda) e^{ijt} \|_S < \varepsilon .$$

Thus $f \varepsilon \overline{P(A)}^S$ . //

## NOTES

Although it is known that the R-synthesis holds for $A^p(G)$ , where $1 \leq p \leq 2$ and $G$ is an infinite compact abelian group, but fails for $L^p(T)$ and $A^r(T)$ where $1 \leq p < 2$ and $2 < r < \infty$ , the theory of closed subalgebras of Segal algebras is far from being complete such as the following natural question is still open: Does the R-synthesis hold for $L^p(T)$ , $2 < p < \infty$?

# CHAPTER V

# HOMOMORPHISMS OF HOMOGENEOUS BANACH ALGEBRAS

In this chapter, we devote ourselves to study the complex homomorphisms, the homomorphisms, and the multipliers of homogeneous Banach algebras through sections 11 and 12.

## 11. COMPLEX HOMOMORPHISMS OF HOMOGENEOUS BANACH ALGEBRAS

This section chiefly contains a characterization of all complex homomorphisms of homogeneous Banach algebras.

We proceed to prove the main result, beginning with some definitions.

11.1. <u>DEFINITIONS</u> Let $B$ be a Banach algebra. $A$ is called a Banach algebra in $B$ if $A$ is a subalgebra of $B$ and $A$ itself is a Banach algebra under some suitable norm.

A complex homomorphism of $B$ is a non-trivial algebraic homomorphism of $B$ into the complex field.

11.2. <u>THEOREM</u> Let $(A, \| \ \|_A)$ be a commutative Banach algebra with the maximal ideal space $m(A)$. If $(I, \| \ \|_I)$ is a Banach algebra in $A$ as well as an ideal in $A$, then the maximal ideal space $m(I)$ of $I$ is canonically homeomorphic to the subspace $m(A) \setminus Z(I)$ of $m(A)$ where $Z(I) = \{h \in m(A): h(I) = 0\}$, the zero set of $I$.

<u>PROOF</u>. Suppose $h \in m(I)$, take $a \in I$ such that $h(a) \neq 0$.

$h^{\#}(x) = h(xa)/h(a)$ for $x \in A$. $h^{\#}$ is independent of the choice of $a$. For $a,b \in I$ such that $h(a) \neq 0$, $h(b) \neq 0$. Then $h(xa)\,h(b) = h(xab) = h(xb)h(a)$ for $x \in A$. Consequently, $h(xa)/h(a) = h(xb)/h(b)$. Moreover, $h^{\#}$ is a (non-zero) complex homomorphism of $A$. This assertion follows from

$$\begin{aligned} h^{\#}(xy) &= h(xya)/h(a) \\ &= h(xya)h(a)/h(a)^2 \\ &= h(xa)h(ya)/h(a)^2 \\ &= h^{\#}(x)h^{\#}(y) \end{aligned}$$

and the linearity of $h^{\#}$, which is obvious. Then consider the map $\#$ of $m(I)$ into $m(A) \setminus Z(I)$ defined by

$$h \to h^{\#}$$

(i) # is 1-1.

This assertion follows from that $h^{\#}$ is independent of a.

(ii) # is onto.

Let $H \in m(A) \setminus Z(I)$ and let $h = H|_I$, clearly, $h \in m(I)$. Moreover $H = h^{\#}$, since, for $x \in A$,

$$\begin{aligned} H(x) &= H(xa)/H(a) \quad \text{for some } a \in I \text{ with } h(a) \neq 0 \\ &= h(xa)/h(a) \\ &= h^{\#}(x) . \end{aligned}$$

(iii) # is continuous.

Suppose $(h_\lambda)$ is a net in $m(I)$ with $h_\lambda \to h$ in the w*-topology for some $h \in m(I)$. Take $a \in I$ with $h(a) = 1$.

Then, for $x \in A$ .

$$\begin{aligned} |h_\lambda^{\#}(x) - h^{\#}(x)| &= |h_\lambda^{\#}(x)h(a) - h^{\#}(x)h(a)| \\ &\leq |h_\lambda^{\#}(x)h(a) - h_\lambda^{\#}(x)h_\lambda(a)| \\ &\quad + |h_\lambda^{\#}(x)h_\lambda(a) - h^{\#}(x)h(a)| \\ &\leq \|x\|_A \, |h(a) - h_\lambda(a)| \\ &\quad + |h_\lambda(xa) - h(xa)| \\ &\to 0 \quad \text{as} \quad \to \infty . \end{aligned}$$

Or, $h_\lambda^{\#} \to h^{\#}$ in the w*-topology.

(iv) Clearly, $\#^{-1}$ is continuous.

Combining (i), (ii), (iii) and (iv), we obtain that # is a homeomorphism of $m(I)$ onto $m(A) \setminus Z(I)$ , which will be called the canonical homeomorphism. In this case, we write

$$m(I) = m(A) \setminus Z(I) . \qquad //$$

The last theorem has a couple of applications:

**11.3. THEOREM** Let $B(G)$ be a homogeneous Banach algebra and $h$ a function on $B(G)$ . Then

(i) $h \in m(B(G))$ if and only if there is a $\gamma \in \Gamma \setminus Z(B(G))$ such that $h(f) = \hat{f}(\gamma)$ for all $f \in B(G)$ .

(ii) Suppose $G$ is compact, then $h \in m(B(G))$ if and only if there exists $\gamma \in \Gamma \cap B(G)$ such that $h(f) = \hat{f}(\gamma)$ for all $f \in B(G)$ .

**PROOF.** (i) Recall that $B(G)$ is a Banach algebra as well as an ideal in $L^1(G)$ , and that $m(L^1(G))$ can be identified with $\Gamma$ under the map $h \to \gamma$ where $\gamma \in \Gamma$ and $h(f) = \hat{f}(\gamma)$ for $f \in L^1(G)$ .

(ii) The assertion follows from $\Gamma \cap B(G) = \Gamma \setminus Z(B(G))$ for the compact $G$ ; Theorem 1.15. //

11.4. THEOREM Let $S(G)$ be a Segal algebra and $I$ a Banach algebra as well as an ideal in $S(G)$ . Then $I$ is dense in $S(G)$ if and only if $m(I)$ is canonically homeomorphic to $m(S(G))$ . In particular, $m(S(G)) = \Gamma$ .

PROOF. Suppose $\bar{I}^S = S(G)$ , then plainly, $Z(I) = Z(S(G))$ . Since $\overline{S(G)}^{L^1} = L^1(G)$ , $Z(S(G)) = Z(L^1(G)) = \phi$ . Hence $Z(I) = \phi$ . By Theorem 11.2, $m(I) = m(S(G)) \setminus Z(I)$ . Or, $m(I) = m(S(G))$. Conversely, if $m(I) = m(S(G))$ , then $Z(I) = \phi$ . By Theorem 3.6, $I$ is an ideal in $L^1(G)$ . We apply Wiener's Tauberian Theorem to get $\bar{I}^{L^1} = L^1(G)$ . But, by Theorem 4.3,

$$\begin{aligned} \bar{I}^S &= \overline{\bar{I}^S}^{L^1} \cap S(G) \\ &= \bar{I}^{L^1} \cap S(G) \\ &= L^1(G) \cap S(G) \\ &= S(G) . \end{aligned}$$

This proves the theorem. //

11.5. THEOREM Let $B$ be a commutative Banach algebra and $I$ a Banach algebra as well as a regular ideal in $B$ . Then

$I = B$ if and only if $m(I) = m(B)$ .

PROOF. It suffices to prove that if $m(I) = m(B)$ , then $I = B$ . Assume $I \subsetneqq B$ , then there is a regular maximal ideal $M$ in $B$ such that $I \subseteqq M \subsetneqq B$ . Take $H \in m(B)$ such that $H(M) = 0$. We conclude that $H(I) = 0$ . This asserts that $m(I)$ can not be the whole of $m(B)$ under the canonical identification. //

A consequence of Theorem 11.5 is as follows: Let $G$ be a non-discrete locally compact abelian group and $M(G)$ the measure algebra. Since $M(G)$ admits an identity, every ideal in $M(G)$ is regular. In particular, $L^1(G)$ is a proper regular ideal in $M(G)$ . By Theorem 11.5, $m(L^1(G)) \neq m(M(G))$ . In other words, for each $\gamma \in \Gamma$ , $\mu \to \hat{\mu}(\gamma)$ determines a complex homomorphism on $M(G)$ but, nevertheless, not every complex homomorphism can be acquired in this way. Loosely speaking, $m(M(G))$ contains $\Gamma$ as a proper subspace.

There is a Banach algebra in $L^1(G)$ , which is not an ideal in $L^1(G)$ , but whose maximal ideal space is bigger than $\Gamma$ . Consider the weight function $w(x) = e^{2|x|}$ defined on the real line $R$ , and the Beurling algebra $L^1_w(R)$ . By the remark after Theorem 5.6, $m(L^1_w(R))$ is the strip $\{z : z$ a complex with $|Re(z)| \leq 2\}$ which contains $R$ as a proper (identified) subspace. Since $R = m(L^1(R))$ , and $L^1_w(R)$ is a Banach algebra in $L^1(G)$ which is not an ideal in $L^1(G)$ .

# 12. HOMOMORPHISMS AND MULTIPLIERS OF HOMOGENEOUS BANACH ALGEBRAS

This section contains the characterizations of endomorphisms and automorphisms of the $A^p(T)$-algebras, $1 \leq p \leq 2$, and the characterization of multipliers of a class of homogeneous Banach algebras.

Boas asked the question: For which sequence of integers $(\alpha(n))_{n=-\infty}^{\infty}$, $\sum a(\alpha(n))e^{int}$ is a Fourier series whenever $\sum a(n)e^{int}$ is a Fourier series?

We may investigate Boas' question in a more general setting: Let $B_1(G_1)$ and $B_2(G_2)$ be two homogeneous Banach algebras where $G_1$ and $G_2$ are two locally compact abelian groups with character groups $\Gamma_1$ and $\Gamma_2$, respectively.

I. If $\Phi : B_1(G_1) \rightarrow B_2(G_2)$ is a homomorphism, then for every $\gamma \in \Gamma_2$, the map $f \rightarrow \widehat{\Phi f}(\gamma)$ is a complex homomorphism (may be trivial) of $B(G_1)$. Let $Y$ be the set of all $\gamma \in \Gamma_2$ for which the homomorphism is not trivial. If $\gamma \in Y$, Theorem 11.3 shows that there is $\alpha(\gamma) \in \Gamma_1$ such that $\widehat{\Phi f}(\gamma) = \hat{f}(\alpha(\gamma))$. Thus each homomorphism $\Phi$ of $B_1(G_1)$ into $B_2(G_2)$ induces a map $\alpha$ of a subset $Y$ of $\Gamma_2$ into $\Gamma_1$, such that

$$\widehat{\Phi f}(\gamma) = \begin{cases} \hat{f}(\alpha(\gamma)) & \text{if } \gamma \in Y \\ 0 & \text{otherwise} \end{cases} \tag{1}$$

for $f \in B_1(G)$ and $\gamma \in \Gamma_2$.

II. Let $Y$ be a subset of $\Gamma_2$, and $\alpha: Y \to \Gamma_1$ a map such that for every $f$ in $B_1(G_1)$ there is $g \in B_2(G)$ satisfying

$$\hat{g}(\gamma) = \begin{cases} \hat{f}(\alpha(\gamma)) & \gamma \in Y \\ 0 & \text{otherwise} \end{cases} \tag{2}$$

If we write $\Phi_\alpha f$ instead of $g$, then, clearly, $\Phi_\alpha$ is a homomorphism of $B_1(G)$ into $B_2(G)$.

The map $\alpha$ described in either I or II is called a carried map of $B_1(G_1)$ into $B_2(G_2)$. We shall abbreviate (1), (2) by $\widehat{\Phi f} = \hat{f}o\alpha$. In case $G_1 = G_2$ and $B_1(G_1) = B_2(G_2)$, $\alpha$ is called a carried map of $B_1(G_1)$. The above remarks show that the characterization of all homomorphisms of $B_1(G_1)$ into $B_2(G_2)$ is equivalent to the characterization of all carried maps of $B_1(G_1)$ into $B_2(G_2)$.

The decisive step in all homomorphism theorems was taken by Cohen ( 1960 ) : Suppose that $G_1$ and $G_2$ are two locally compact abelian groups with character groups $\Gamma_1$ and $\Gamma_2$, respectively. If $E$ is a coset in $\Gamma_2$ and $\alpha$ is a continuous map of $E$ into $\Gamma_1$ which satisfies the identity

$$\alpha(\gamma + \gamma' - \gamma'') = \alpha(\gamma) + \alpha(\gamma') - \alpha(\gamma'') \qquad (\gamma, \gamma', \gamma'' \in E)$$

(note that $E + E - E \subset E$) then $\alpha$ is said to be affine. Suppose that

(i) $S_1, \cdots, S_n$ are pairwise disjoint sets belonging to the coset-ring of $\Gamma_2$ ,

(ii) each $S_i$ is contained in an open coset $K_i$ in $\Gamma_2$,

(iii) for each $i$ , $\alpha_i$ is an affine map of $K_i$ into $\Gamma_1$,

(iv) $\alpha$ is the map of $Y = S_1 \cup \cdots \cup S_n$ into $\Gamma_1$ which coincides on $S_i$ with $\alpha_i$ .

Then $\alpha$ is said to be a piecewise affine map of $Y$ into $\Gamma_1$ .

12.1. THEOREM [Cohen]. If $\Phi$ is a homomorphism of $L^1(G_1)$ into $M(G_2)$ , then $\widehat{\phi f} = \hat{f} \circ \alpha$ , where $\alpha$ is a piecewise affine map of $Y$ into $\Gamma_1$ and $Y$ belongs to the coset-ring of $\Gamma_2$ .

Conversely, if $Y$ belongs to the coset-ring of $\Gamma_2$ and if $\alpha$ is a piecewise affine map of $Y$ into $\Gamma_1$ , then $\mu \to \phi\mu$ is a homomorphism of $M(G_1)$ into $M(G_2)$ where $\widehat{\phi\mu} = \hat{\mu} \circ \alpha$ .

Rudin restated the preceding theorem in a quite concrete form: Suppose $Y$ is a subset of $Z$ and $\alpha$ maps $Y$ into $Z$ . For which $Y$ and $\alpha$, is it true that

$$\sum_{n \in Y} c(\alpha(n)) e^{int}$$

is a Fourier series (of a function in $L^1(T)$) whenever the series

$$\sum_{n \in Z} c(n)\, e^{int}$$

if a Fourier series? Or for which map $\alpha$ is a carried map of $L^1(T)$ .

12.2. <u>THEOREM</u> [Rudin (1962 ,p.95)] $\alpha : Y \to Z$ is a carried map of $L^1(T)$ if and only if there are a positive integer $q$ and a map $\beta$ of $Z$ into $Z$ with the following properties:

(i) If $A_1,\cdots,A_q$ are the residue classes modulo $q$, then $Y = S_1 \cup \cdots \cup S_n$ where each $S_i$ is either finite or contained in some $A_j$ from which it differs by a finite set, and the sets $S_i$ are pairwise disjoint.

(ii) $\alpha(n) = \beta(n)$ for all $n \in Y$, with possibly finitely many exceptions.

(iii) $\beta(n + q) \neq \beta(n)$ for all $n \in Z$.

(iv) $\beta(n + q) + \beta(n - q) = 2\beta(n)$ for all $n \in Z$.

The following results reveal that every carried map $\alpha$ of $L^1(T)$ satisfies the property: there is a constant $C > 0$ such that

$$\alpha^{-1}(n)^{\#} \leq C \quad (n \in Z) \tag{3}$$

where $\alpha^{-1}(n)^{\#}$ denotes the cardinal number of $\alpha^{-1}(n)$. But not every map $\alpha$ satisfying (3) is a carried map of $L^1(T)$.

12.3. <u>THEOREM</u>

(i) For a carried map $\alpha$ of $L^1(T)$, there is a constant $C > 0$ such that $\alpha^{-1}(n)^{\#} \leq C$ for all $n \in Z$.

(ii) There exists a map $\alpha : Z \to Z$ which is not a carried map of $L^1(T)$ but there is $C > 0$, $\alpha^{-1}(n)^{\#} \leq C$ for all $n \in Z$.

<u>PROOF</u>. (i) By Theorem 12.2, there are a positive integer $q$ and a map $\beta$ of $Z$ into $Z$ with the properties (a) ~ (d) of that theorem. In order to show that $\alpha^{-1}(n)^{\#} \leq C_1$ for all $n \in Z$, it suffices to show that $\beta^{-1}(n)^{\#} \leq C$ for all $n \in Z$. For $n \in Z$, we have

$$\beta(n + q) - \beta(n) = \beta(n) - \beta(n - q) .$$

Consequently the map $\beta(n + q) - \beta(n)$ is constant, $d_i$, say, on each residue class $A_1$ modulo $q$, $i = 1,2,\cdots,q$. If $i = 1,2,\cdots,q$, then

$$\begin{aligned}\beta(i+kq) - \beta(i) &= \beta(i+kq) - \beta(i+(k-1)q) + \beta(i+(k-1)q) \\ &\quad - \beta(i+(k-2)q) + \cdots + \beta(i+q) - \beta(i) \\ &= kd_i .\end{aligned}$$

Or,

$$\beta(i+kq) = \beta(i) + kd_i .$$

By Theorem 12.2 (iii), $d_i \neq 0$, so the restriction of $\beta$ on each $A_i$ is a 1-1 map. We conclude that $\beta^{-1}(n)^{\#} \leq q$ for all $n \in Z$.

(ii) Define a map $\alpha$ of $Z$ into $Z$ by $\alpha(n) = |n|$. Then $\alpha^{-1}(n)^{\#} \leq 2$ for all $n \in Z$. We claim that $\alpha$ is not a carried map of $L^1(T)$. Let $\beta$ be any map of $Z$ into $Z$ such that $\beta(n) = \alpha(n)$ for all, with possibly finitely many exceptions. It is clear that $\alpha$ does not satisfy the condition (iv) of Theorem 12.2. For instance, for any $q > 0$, let $n = 0$ then $|n + q| + |n - q| = 2|q| \neq 0 = 2|n|$. Next, let

$F = \{n \in Z : \beta(n) \neq \alpha(n)\} \neq \phi$ , and

$$m = \max_{n \in F} n$$

$$\ell = \min_{n \in F} n .$$

Suppose $q$ is an arbitrary positive integer. For $m \geq 0$ , we have

$$\begin{aligned} \beta(m+2q) + \beta(m) &= \alpha(m + 2q) + \beta(m) \\ &= |m + 2q| + \beta(m) \\ &\neq |m + 2q| + |m| \\ &= 2|m + q| \\ &= 2\beta(m + q) . \end{aligned}$$

In case $m < 0$ , we have

$$\begin{aligned} \beta(\ell) + \beta(\ell-2q) &= \beta(\ell) + |\ell - 2q| \\ &\neq \alpha(\ell) + |\ell - 2q| \\ &= |\ell| + |\ell - 2q| \\ &= 2|\ell - q| \\ &= 2\beta(\ell - q) . \end{aligned}$$

Consequently, (iv) of Theorem 12.2 is not fulfilled. //

12.4. <u>THEOREM</u> Let $1 \leq p \leq 2$ , $Y$ a subset of $Z$ , and $\alpha$ a map of $Y$ into $Z$ . Then $\alpha$ is a carried map of $A^p(T)$ if and only if the set $\{(\alpha^{-1}(n)^{\#} : n \in Z\}$ is bounded above.

<u>PROOF.</u> "Only-if-part" Suppose $\alpha$ is a carried map of $A^p(T)$. We proceed to prove that the set $\{\alpha^{-1}(n)^{\#} : n \in Z\}$ is bounded

above, beginning with remark that $\alpha^{-1}(n)^{\#}$ is finite for every $n \varepsilon Z$. For otherwise, if $\alpha^{-1}(k)$ is an infinite set for some $k \varepsilon Z$. Put $f(t) = e^{ikt}$. Then, plainly, $f \varepsilon A^p(T)$, and

$$\sum_{n\varepsilon Y} |\hat{f}(\alpha(n))|^p \geq \sum_{n\varepsilon\alpha^{-1}(k)} |\hat{f}(\alpha(n))|^p = \infty .$$

So $\alpha$ is not a carried map of $A^p(T)$. The assertion of this part is trivial if $Y$ is a finite subset of $Z$. Suppose $Y$ is infinite and $\{\alpha^{-1}(n)^{\#} : n \varepsilon Z\}$ is not bounded above. Take a sequence of integers $(n_j)_{j=1}^{\infty}$ such that

$$n_i \neq n_j \quad \text{if} \quad i \neq j ,$$

and

$$\alpha^{-1}(n_j)^{\#} \geq j^j \quad j = 1,2,\cdots\cdots .$$

Since

$$\sum_{j=1}^{\infty} \frac{\|e^{in_jt}\|_{L^1}}{j^2} < \infty$$

there is $f \varepsilon L^1(T)$ such that

$$f = \sum_{j=1}^{\infty} \frac{e^{in_jt}}{j^2} .$$

Consequently,

$$\hat{f}(n) = \begin{cases} \frac{1}{j^2} & n = n_j \\ 0 & \text{otherwise} \end{cases}$$

Then, plainly, $f \varepsilon A^p(T)$ for $1 \leq p \leq 2$. But $\hat{f}o\alpha \notin A^p(T)$ since

$$\sum_{n\varepsilon Z}|\hat{f}(\alpha(n))|^p \geq \sum_{j=1}^{\infty}\sum_{n\varepsilon\alpha^{-1}(n_j)}|\hat{f}(\alpha(n))|^p$$

$$\geq \sum_{j=1}^{\infty} j^j|\hat{f}(n_j)|^p$$

$$= \sum_{j=1}^{\infty} j^j \frac{1}{j^{2p}}$$

$$= \infty .$$

It turns out that $\alpha$ is not a carried map of $A^p(T)$ .

"If-part" Suppose the set $\{\alpha^{-1}(n)^{\#} : n \varepsilon Z\}$ is bounded by $C > 0$ . We claim that there exists $g \varepsilon A^p(T)$ such that $\hat{g} = \hat{f} \circ \alpha$ . In fact, if $f \varepsilon A^p(T)$ , then

$$\sum_{n\varepsilon Y}|\hat{f}(\alpha(n))|^p \leq C \sum_{n\varepsilon Z}|f(n)|^p < \infty .$$

In particular,

$$\sum_{n\varepsilon Y}|\hat{f}(\alpha(n))|^2 < \infty .$$

By the Plancherel Theorem, there exists $g \varepsilon L^2(T)$ such that

$$\hat{g}(n) = \begin{cases} \hat{f}(\alpha(n)) & \text{if } n \varepsilon Y \\ 0 & \text{otherwise} \end{cases}$$

Moreover

$$\sum_{n\varepsilon Z}|\hat{g}(n)|^p = \sum_{n\varepsilon Y}|\hat{g}(n)|^p$$

$$= \sum_{n\varepsilon Y}|\hat{f}(\alpha(n))|^p$$

$$< \infty .$$

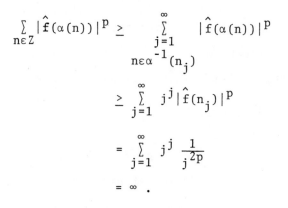

Consequently $g \in A^p(T)$ , or $\alpha$ is a carried map of $A^p(T)$. //

12.5. <u>REMARK</u> Reviewing the "only-if-part" proof of the preceding theorem, we conclude that if $\alpha$ is a carried map of $A^p(T)$ where $1 \leq p < \infty$ , then the set $\{\alpha^{-1}(n)^{\#} : n \in Z\}$ is bounded above.

12.6. <u>THEOREM</u> Every carried map of $L^1(T)$ is a carried map of $A^p(T)$ , $1 \leq p < \infty$ . However, there is a carried map of $A^p(T)$ , $1 \leq p < 2$ , which is not a carried map of $L^1(T)$ .

<u>PROOF.</u> Let $\alpha$ be a carried map of $L^1(T)$ . By theorem 12.3 (i), there is a constant $C > 0$ such that

$$\alpha^{-1}(n)^{\#} \leq C \quad \text{for all} \quad n \in Z .$$

For $f \in A^p(T)$ , take $g \in L^1(T)$ such that

$$\hat{g}(n) = \begin{cases} \hat{f}(\alpha(n)) & \text{if} \quad n \in Y \\ 0 & \text{otherwise} \end{cases}$$

Now

$$\begin{aligned} \sum_{n \in Z} |\hat{g}(n)|^p &= \sum_{n \in Y} |\hat{f}(\alpha(n))|^p \\ &\leq C \sum_{n \in Z} |\hat{f}(n)|^p \\ &< \infty . \end{aligned}$$

Hence $g \in A^p(T)$ and then $\alpha$ is a carried map of $A^p(T)$ .

The second part of the theorem follows from Theorems 12.3 (ii), and 12.4. //

As a consequence of Theorem 12.4, a permutation on $Z$ is a carried map of $A^p(T)$, $1 \leq p \leq 2$. Moreover,

12.7. THEOREM Let $1 \leq p \leq 2$. If $\alpha$ is a permutation on $Z$, then $\Phi$ is an automorphism of $A^p(T)$ where $\widehat{\Phi f} = \hat{f} \circ \alpha$. Conversely if $\Phi$ is an automorphism of $A^p(T)$, then there is a permutation $\alpha$ on $Z$ such that $\widehat{\Phi f} = \hat{f} \circ \alpha$.

PROOF. Suppose $\alpha$ is a permutation on $Z$, and $\alpha^{-1}$ is the inverse map of $\alpha$. By Theorem 12.4, $\alpha$ and $\alpha^{-1}$ induce endomorphisms of $A^p(T)$, say, $\Phi_\alpha$ and $\Phi_{\alpha^{-1}}$, respectively. Then

$$\widehat{\Phi_\alpha f} = \hat{f} \circ \alpha$$

and

$$\widehat{\Phi_{\alpha^{-1}} f} = \hat{f} \circ \alpha^{-1}$$

for $f \in A^p(T)$. Plainly, we have $\Phi_\alpha \circ \Phi_{\alpha^{-1}}$ and $\Phi_{\alpha^{-1}} \circ \Phi_\alpha$ are the identity map of $A^p(T)$. Consequently, $\Phi_\alpha$ is 1-1 and onto. Thus $\Phi_\alpha$ is an automorphism of $A^p(T)$.

Conversely, suppose $\Phi$ is an automorphism of $A^p(T)$. Let $\Phi^{-1}$ be the inverse map of $\Phi$. Then $\Phi^{-1}$ is also an automorphism of $A^p(T)$. Let $\alpha$ and $\beta$ be the carried maps associated with $\Phi$ and $\Phi^{-1}$, respectively. That is, $\widehat{\Phi f} = \hat{f} \circ \alpha$ and $\widehat{\Phi^{-1} f} = \hat{f} \circ \beta$. Since $\Phi \circ \Phi^{-1}$ and $\Phi^{-1} \circ \Phi$ are the identity map of $A^p(T)$, $\alpha \circ \beta$, and $\beta \circ \alpha$ are the identity map of $Z$. Thus $\alpha$ is a permutation. //

We wish in the final of this section to present some

results on the multipliers of homogeneous Banach algebras. There is a good reason to consider such a subject in this stage. By a multiplier of a homogeneous Banach Algebra $B(G)$ we mean a bounded linear operator $T : B(G) \to B(G)$ such that $T$ commutes with translations or equivalently, $T(f * g) = f * Tg$ for all $f, g \in B(G)$. Denote the set of all multipliers of $B(G)$ by $M(B(G))$, then, by Ching-Wong (1967), $M(B(G))$ form a closed commutative subalgebra of the Banach algebra of all bounded linear operators on $B(G)$ and is called the multiplier algebra of $B(G)$. Suppose $B_1(G_1)$ and $B_2(G_2)$ are two homogeneous Banach algebras and $T : B_1(G_1) \to B_2(G_2)$ is an algebraic isomorphism (onto), then $U_2 \to T^{-1}U_2T$ is an 1-1 correspondence between the multiplier algebras $M(B_2(G_2))$ and $M(B_1(G_1))$.

Before exposing the multipliers of general homogeneous Banach algebras, we describe the multipliers of some special Segal algebras in which the theory shows very diverse.

(i) [Figà-Talamanca-Gaudry (1970)] For any non-compact locally compact abelian group $G$, let $S(G)$ by any one of the Segal algebras $L^1(G)$, $A^p(G)$, $L^1 \cap L^p(G)$, and $L^1 \cap C_0(G)$, then every multiplier for $S(G)$ is given by

$$Tf = f * \mu \quad (f \in S(G))$$

where $\mu \in M(G)$.

(ii) [Hewitt-Ross (1970, pp.384-386)] Every multiplier for $L^1(T)$ or $L^2(T)$ is given by

$$Tf = f * \mu$$

where $\mu \in M(T)$ . But for the case of $p \neq 2$ , $1 < p < \infty$ , the multipliers of $L^p(T)$ are as follows : Let $I_p(T)$ consist of all functions $h$ in $C(T)$ that can be written in at least one way as $\sum_{k=1}^{\infty} f_k * g_k$ , where each $f_k$ belongs to $L^p(T)$, each $g_k \in L^q(T)$ , $\frac{1}{p} + \frac{1}{q} = 1$ , and $\sum_{k=1}^{\infty} \|f_k\|_{L^p} \|g_k\|_{L^q} < \infty$ . For $h \in I_p(T)$ , define

$$\|h\|_{I_p} = \inf \{ \sum_{k=1}^{\infty} \|f_k\|_{L^p} \|g_k\|_{L^q} : h = \sum_{k=1}^{\infty} f_k * g_h \} .$$

With this norm, $I_p(T)$ forms a Banach space. Denote the dual space of $I_p(T)$ by $I_p{}^*(T)$ . It turns out $M(T) \subsetneqq I_p{}^*(T)$ and every multiplier for $L^p(T)$ is given by

$$Tf = f * \phi$$

where $\phi \in I_p{}^*(T)$ .

In the followings we attempt to characterize the multipliers of a class of homogeneous Banach algebras : Let $M_\Delta(G)$ be the set of all $\mu \in M(G)$ such that $\hat{\mu}(\Delta^c) = 0$ , where $\Delta$ is any set in $\Gamma$ . Then $M_\Delta(G)$ is a closed subalgebra of $M(G)$ .

12.8. <u>THEOREM</u> Suppose that $G$ is a locally compact abelian group and $B(G)$ is a homogeneous Banach algebra. If $B(G)$ admits a bounded approximate identity and $Z(B(G))$ belongs to the coset ring of $\Gamma$ , then $M(B(G))$ is isomorphic to $M_{Z(B(G))^c}(G)$ . More precisely, for every $\mu \in M_{Z(B(G))^c}(G)$ , $f \to \mu * f$ is a multiplier of $B(G)$ , and for every multiplier $U$ of $B(G)$ , there exists uniquely a $\mu \in M_{Z(B(G))^c}(G)$ such that $Uf = \mu * f$ for all $f \in B(G)$ and that $U \to \mu$ is an isomorphism of $M(B(G))$ onto $M_{Z(B(G))^c}(G)$ .

PROOF. Suppose $\mu \in M_{Z(B(G))^c}(G)$ , then, plainly, $f \to \mu * f$ is a multiplier of $B(G)$ . Conversely, assume that $U$ is a multiplier of $B(G)$ , and that $(k_\lambda)$ is an approximate identity in $B(G)$ bounded by $C$ . For each $\lambda$ ,

$$\begin{aligned} \|Uk_\lambda\|_M &= \|Uk_\lambda\|_{L^1} \\ &\le \|Uk_\lambda\|_B \\ &\le \|U\| \, \|k_\lambda\|_B \\ &\le C\,\|U\| \end{aligned}$$

therefore, by the Alaoglu Theorem, there is $\nu \in M(G)$ and a subnet of $(k_\lambda)$ , still denoted by $(k_\lambda)$ such that

$$Uk_\lambda \to \nu \text{ in the w*-topology .}$$

Consequently

$$Uk_\lambda * g(x) \to \nu * g(x) \quad \text{everywhere} \qquad (1)$$

for $g \in B \cap C_0(G)$ .

On the other hand, since $k_\lambda * g \to g$ in the B-norm for each $g \in B \cap C_0(G)$ , $(Uk_\lambda) * g = U(k_\lambda * g) \to Ug$ in the B-norm.

There exists a subnet $(Uk_{\lambda_\beta})$ of $(Uk_\lambda)$ such that

$$Uk_{\lambda_\beta} * g \rightarrow Ug \quad \text{almost everywhere} . \qquad (2)$$

Conbining (1) and (2), we get $Ug = v * g$ for all $g \in B \cap C_o(G)$. By Theorem 3.7 (ii) (with $E = L^1 \cap C_o(G)$) , $B \cap C_o(G)$ is dense in $B(G)$ , it turns out

$$Uf = v * f \quad (f \in B(G)) .$$

Since $Z(B(G))$ belongs to the coset ring of $\Gamma$ , by Theorem 1.13, there is $\zeta \in M(G)$ such that $\hat{\zeta}(Z(B(G))) = 0$ and $\hat{\zeta}(Z(B(G)^c)) = 1$ . Put $\mu = \zeta * v$ , then $\mu \in M_{Z(B(G))c}(G)$ and

$$\begin{aligned} Uf &= v * f \\ &= \mu * f \quad (f \in B(G)) . \end{aligned}$$

Then, plainly, $U \rightarrow \mu$ , is an isomorphism of $M(B(G))$ onto $M_{Z(B(G)^c}(G)$ . This proves the assertion. //

The following results about Segal algebras was due to Unni and Burnham-Goldberg, respectively.

12.9. <u>THEOREM</u> [Unni (1974)] Let $S(G)$ be a Segal algebra. If $T$ is a multiplier on $S(G)$ , then there exists a unique pseudomeasure $\sigma$ such that

$$Tf = \sigma * f \quad \text{for each} \quad f \in S(G) .$$

<u>PROOF</u>. If $T$ is a multiplier on $S(G)$ , then by a standard

argument, we have a continuous bounded function $\Phi$ defined on $\Gamma$ such that

$$\widehat{Tf} = \Phi\hat{f} \quad (f \in S(G)) .$$

Hence Tf is a continuous function for each $f \in P(G)$ . Define $L(f) = Tf(o)$ then

$$\begin{aligned} |L(f)| &= |Tf(o)| \leq \|Tf\|_\infty \leq \|\widehat{Tf}\|_{L^1} \\ &= \|\phi\hat{f}\|_{L^1} \leq \|\Phi\|_\infty \|\hat{f}\|_{L^1} \\ &= \|\phi\|_\infty \|f\|_{A(G)} . \end{aligned}$$

Thus L is a continuous linear functional on P(G) which is dense in A(G) . Extended L as a continuous linear functional on A(G) . Therefore there is a unique pseudomeasure $\sigma$ with

$$Tf(o) = L(f) = \langle f,\sigma\rangle \quad (f \in S(G))$$

or

$$Tf_x(o) = \langle f_x,\sigma\rangle .$$

That is,

$$\begin{aligned} Tf(-x) &= \int f_x(y)\sigma(-y)dy \\ &= \int f(y - x) \ (-y)dy \\ &= \sigma * f(-x) \qquad \text{a.e.} \end{aligned}$$

or

$$Tf = \sigma * f .$$

12.10. <u>THEOREM</u> [Burnham-Goldberg (1975)] Let $S = S(G)$ be a Segal algebra with relative completion $\tilde{S}$ . If $(L^1,S) \subset L^1$, then $(L^1,S) = \tilde{S}$ . In this case if for $f \in \tilde{S}$ , define $T_f \in (L^1,S)$ by

$$T_f(g) = f * g \quad (g \in L^1)$$

then $f \leftrightarrow T_f$ is an isometric isomorphism of $\tilde{S}$ onto $(L^1,S)$.

<u>PROOF.</u> (i) $(L^1,S) \subset L^1 \Longrightarrow (L^1,S) \subset \tilde{S}$ .

Let $\mu \in (L^1,S)$ . Take $K > 0$ with

$$\|\mu * f\|_S \leq K\|f\|_1 \quad (f \in L^1) .$$

Let $(e_\alpha) \subset S$ be a common approximate identity with $\|e_\alpha\|_1 = 1$ for all $\alpha$ . We obtain

$$\|\mu * e_\alpha\|_S \leq K \text{ for all } \alpha .$$

Now $\mu \in L^1$ , so, by Theorem 6.3, $\mu \in \tilde{S}$ .

(ii) $\tilde{S} \subset (L^1,S)$

If $\mu \in \tilde{S}$ , then $\mu \in (L^1,\tilde{S})$ since $\tilde{S}$ is an ideal of $L^1$ . Take $H > 0$ with

$$|||\mu * f||| \leq H\|f\|_1 \quad (f \in L^1) .$$

But

$$\begin{aligned} |||(\mu * f)_a - \mu * f||| &= |||\mu * f_a - \mu * f||| \\ &\leq H \|f_a - f\|_1 \\ &\to 0 \text{ as } a \to 0 . \end{aligned}$$

By Theorem 6.8 (ii), $\mu * f \in S$ , and then

$$\|\mu * f\|_S = |||\mu * f||| \leq H$$

Thus $\mu \in (L^1, S)$ .

(iii) $||Tf|| = |||f|||$ $(f \in \tilde{S})$ .

For $f \in \tilde{S}$ . Then

$$||f * e_\alpha|| = ||T_f e_\alpha||_S \leq ||T_f|| \, ||e_\alpha||_1$$
$$= ||T_f|| \quad \text{for all } \alpha .$$

Thus

$$|||f||| \leq ||T_f|| .$$

In the other direction, if $f \in \tilde{S}$ , $g \in L^1$ then $f * g \in S$ since $\tilde{S} \subset (L^1, S)$ . Then

$$||f * g * e_\alpha||_S \leq ||g||_1 ||f * e_\alpha||_S .$$

Taking the sup over $\alpha$ we have

$$|||f * g||| \leq ||g||_1 \, |||f||| .$$

But $|||f * g||| = ||f * g||_S$ since $f * g \in S$ . Hence

$$||f * g||_S \leq ||g||_1 \, |||f|||$$

or

$$||T_f \, g||_S \leq ||g||_1 \, |||f||| \quad (g \in L^1) .$$

This implies

$$||T_f|| \leq |||f||| .$$

Together with (i), (ii) and (iii), we completes the proof.

# TABLE OF GROUP ALGEBRAS

| SYMBOL | THE BANACH ALGEBRA OF ALL | CLASSIFICATION AND REFERENCE |
|---|---|---|
| $C^k(T)$ , $1 \leq k < \infty$ , $T$ the circle group | functions with $k$ continuous derivatives on $T$ under the norm $\|f\| = \sup_{0 \leq j \leq k} \|f^{(j)}\|_\infty$ | non-character Segal algebra, FP-algebra |
| $L^{(k)}(T)$ , $1 \leq k < \infty$ | functions $f$ such that $f^{(j)}$ , $j = 0,1,\cdots,k-1$, are absolutely continuous and are in $L^1(T)$ , under the norm $\|f\| = \sup_{0 \leq j \leq k} \|f^{(j)}\|_{L^1}$ | non-character Segal algebra, FP-algebra |
| $BV(T)$ | functions $f$ which are continuous on $T$ with bounded total variation $V_0^{2\pi} f$ , under the norm $\|f\| = \|f\|_{L^1} + V_0^{2\pi} f$ | semi-homogeneous Banach algebra, FP-algebra |
| $Lip_\alpha(T)$ , $0 < \alpha \leq 1$ | functions $f$ of Lipschitz class $\alpha$ | semi-homogeneous Banach algebra, FP-algebra |

| | | |
|---|---|---|
| D(T) | functions $f$ in $L^1(T)$ with $\|f - D_N * f\|_{L^1} \to 0$, where $(D_N)$ is the Dirichlet kernel, under the norm $\|f\| = \sup_{n\geq 1} \|D_n * f\|_{L^1}$ | Segal algebra, P-algebra Burnham (1975,4) |
| E(T) | functions $f$ in $L^1(T)$ for which $\|f\| = \sup_{n\geq 1} \|D_n * f\|_{L^1} < \infty$ under the norm $\|f\|$ | semi-homogeneous Banach algebra, P-algebra |
| C(G), G an infinite compact abelian group | continuous functions on G under the supremum norm | character Segal algebra, FP-algebra |
| $L^p(G), 1 \leq p < \infty$, G an infinite compact abelian group with normalized Haar measure dx | measurable functions f on G with $\|f\| = (\int_G \|f(x)\|^p dx)^{\frac{1}{p}} < \infty$ under the norm $\|f\|$ | character Segal algebra, FP-algebra |
| $L^\infty(G)$, G an infinite compact abelian group | essentially bounded functions on G under the essential supremum norm | semi-homogeneous Banach algebra, FP-algebra |

| | | |
|---|---|---|
| $L^{(k)}(R), 1 \leq p < \infty$, R the additive group of all real numbers | functions f in $L^1(R)$ such that $f^{(j)}$, $j=0,1,\cdots,k-1$, are absolutely continuous on R and are in $L^1(R)$ under the norm $\|f\| = \sup_{0\leq j\leq k} \|f^{(j)}\|_{L^1}$ | non-character Segal algebra, FP-algebra |
| $L_{BV}(R)$ | functions f in $L^1(R)$ which are bounded variation on R , under the norm $\|f\| = \|f\|_{L^1} + V_R f$ | semi-homogeneous Banach algebra, FP-algebra Burnham (1975,4) |
| $F(R)$ | functions f in $L^1(R)$ with $\lim_{n\to\infty} \hat{f}(n)\log n = 0$ , under the norm $\|f\| = \|f\|_{L^1} + \sup_{n\geq 1} \|\hat{f}(n)\|\log n$ | Segal algebra, non-F algebra Goldberg (1974) |
| $W(G)$, G a non-discrete locally compact abelian group having a discrete subgroup H such that G/H is compact (There will then exist a compact set K of | functions f in $C(G)$ with $\|f\| = \sup_{u\in G} \sum_{h\in H} \max_{x\in K} \|\hat{f}(uhx)\| < \infty$ under the norm $\|f\|$ | character Segal algebra, FP-algebra Wiener (1932) Ditkin (1939) Goldberg (1967) Wang (1972) |

| | | |
|---|---|---|
| measure 1 in $G$ such that $G = HK$). For example $G = R$, $H =$ integers, $K = [0,2\pi]$ Haar measure $= \frac{1}{2\pi} \cdot$ Lebesgue measure | | |
| $W_\infty(G)$, $G$ as in $W(G)$ | functions $f$ in $L^\infty(G)$ with $\|f\| = \sup_{u\varepsilon G} \sum_{h\varepsilon H} \operatorname{ess\,sup}_{x\varepsilon K} \|f(uhx)\| < \infty$ under the norm $\|f\|$ | semi-homogeneous Banach algebra, FP-algebra Burnham (1975,4) |
| $L^1 \cap C_o(G)$, $G$ a non-discrete locally compact abelian group | continuous functions $f$ in $L^1(G)$ which vanish at $\infty$ under the norm $\|f\| = \|f\|_{L^1} + \|f\|_\infty$ | character Segal algebra, FP-algebra |
| $L^1 \cap L^p(G)$, $G$ a non-discrete locally compact abelian group | functions $f$ in $L^1(G)$ and $L^p(G)$ under the $\|f\| = \|f\|_{L^1} + \|f\|_{L^p}$ | character Segal algebra, FP-algebra |

| | | |
|---|---|---|
| $L^1 \cap L^\infty(G)$ , G a non-discrete locally compact abelian group | functions f in $L^1(G)$ and $L^\infty(G)$ under the norm $\|\|f\|\| = \|\|f\|\|_{L^1} + \|\|f\|\|_\infty$ | semi-homogeneous Banach algebra, FP-algebra |
| $A^p(\mu), 1 \leq p < \infty$, G a non-discrete locally compact abelian group with character group $\Gamma$ and $\mu$ a positive unbounded regular measure on $\Gamma$ | functions f in $L^1(G)$ whose Fourier transforms $\hat{f}$ belong to $L^p(\mu)$ under the norm $\|\|f\|\| = \|\|f\|\|_{L^1} + \|\|\hat{f}\|\|_{L^p(\mu)}$ | Segal algebra, character if $\mu$ is Haar, $F^\mu P^\mu$-algebra Larsen-Liu-Wang (1964) |
| $\mu * L^1(G)$ , $\mu \in M(G)$ | functions $\mu * f$ where $f \in L^1(G)$ under the norm $\|\|\mu * f\|\| = \|\|f\|\|_{L^1}$ | Homogeneous Banach algebra, Segal algebra if $\hat{\mu}$ is never vanished |
| $S_f$ , $f \in L^1(G)$ | functions g in $L^1(G)$ with $f * g \in C_0(G)$ under the norm $\|\|g\|\| = \|\|g\|\|_{L^1} + \|\|f * g\|\|_\infty$ | Homogeneous Banach algebra, Segal algebra if $\hat{f}$ is never vanished |

| | | |
|---|---|---|
| | | Johnson (1973) |
| $L^1_\Delta(G)$, $\Delta \subset \Gamma$ where $\Gamma$ is the character group of $G$ | functions $f$ in $L^1(G)$ with $\hat{f}(\Delta^c) = 0$ under the $L^1$-norm | Homogeneous Banach algebra, non-Segal if $\Delta \neq \phi$ |
| $B_1^{(\xi, B_2)}(G)$, $B_1(G)$ and $B_2(G)$ are homogeneous Banach algebras, while $\xi$ is a continuous function on $\Gamma$, the character group of $G$ | functions $f$ in $B_1(G)$ with $\xi\hat{f} \in \widehat{B_2(G)}$ under the norm $\|f\| = \|f\|_{B_1} + \|g\|_{B_2}$ where $\hat{g} = \xi\hat{f}$, $g \in B_2(G)$ | Homogeneous Banach algebra, Segal if $B_1$, $B_2$ are Segal and $\xi\widehat{P(L^1)} \subset \widehat{P(L^1)}$ |

## BIBLIOGRAPHY


Altman, M.: [1] (1971) Factorization dans les algèbres de Banach, C. R. Acad. Sc. Paris, 272 (1971), 1388-1389.

[2] (1972) Contracteurs dans des algèbres de Banach, C. R. Acad. Sc. Paris, 274 (1972), 399-400.

[3] (1972) Infinite products and factorization in Banach algebras, Boll. Un. Math. Ital. (4), 5 (1972), 217-229.

[4] (1973) Contractors, Approximate identities and factorization in Banach algebras, Pacific J. of Math. 48 (1973), #2, 323-334.

[5] (1975) A generalization and the converse of Cohen's factorization theorem, Duke Math. J., 42 (1975), 105-110.

Bennett, C. and Gilbert, J. E. (1972): Homogeneous Algebras on the Circle: I, II Ann. Inst., Fourier, Grenoble. 22, 3 (1972), 1-19, 21-50.

Beurling, A. and Helson, H. (1953): Fourier-Stieltjes Transforms with Bounded Powers, Math. Scand, 1 (1953), 120-126.

Bourbaki, N. (1952): Integration, Chapters I-IV, Hermann and Cie, Paris (1952).

Burnham, J. T.: [1] (1972) Closed ideals in subalgebras of Banach algebras, Proc. Amer. Math. Soc., 32 (1972), 551-555.

[2] (1972) Nonfactorization in subsets of the measure algebras, Proc. Amer. Math. Soc., 35 (1972), 104-106.

[3] (1974) Segal algebras and dense ideals in Banach algebras, Springer-Verlag, Lecture Notes in Math., 399 (1974), 33-58.

[4] (1975) The relative completion of an A-Segal algebra is closed. Proc. Amer. Math. Soc., 48 (1975), 1-7.

[5] (1975) Closed ideals in subalgebras of Banach algebras II, Ditkin's Condition, To appear Montrel Für Math.

Burnham, J. T. and Goldberg, R. R.: [1] (1973) Basic properties of Segal algebras, J. of Math. Analy. and Appli., 42 (1973), 323-329.

[2] (1974) The convolution theorems of Dieudonné, Acta Sci. Math., 36 (1974), 1-3.

[3] (1975) Multipliers from $L^1(G)$ to a Segal algebras, Bull. Inst. Math. Acad. Sinica. 2 (1974), 153-164.

Burnham, J. T. and Krogstad H., and Larsen R.: Multipliers and the Hilbert distribution, To appear Nanta Math.

Cheng, C. S. (1973): Segal algebras and the multiplication by continuous characters, Chinese J. of Math., 1 (1973), 175-181.

Ching, W. M. and Wong, J. S. W. (1967): Multipliers and H* algebras. Paci. J. Math., 22 (1967), 387-395.

Cigler, J. (1969): Normed ideals in $L^1(G)$ . Nederl. Akad, Wetensch. Indag. Math., 31 (1969), 273-282.

Cohen, P. J.: [1] (1959) Factorization in group algebras. Duke Math. J. 26 (1959), 199-205.

[2] (1960) On homomorphisms of group algebras, Amer. J. Math. 82 (1960), 213-226.

Curtis, P. C. Jr. and Figà-Talamanca, A. (1966): Factorization theorems for Banach algebras, Function algebras, Atlanta: Scott, Foresman and Company (1966), 169-185.

Dietrich, W. (1972): On the ideal structure of Banach algebras. Trans. Amer. Math. Soc. 169 (1972), 59-74.

Ditkin, V. A. (1939): Study of the structure of ideals in certain normed rings. Učenye Zapiski Moskov. Gos. Univ. Matematika 30 (1939), 83-130.

Dugundji, J. (1966): Topology, Allyn and Bacon, Inc., Boston, Massachusetts (1966).

Dunford, B.: Segal algebras and left normed ideals, Proc. London Math. Soc., To appear.

Edwards, R. E.: [1] (1965) Approximation by convolutions. Pacific J. Math. 15 (1965), 85-95.

[2] (1967) Fourier series; a modern introduction, Vol. I, II, Holt, Rinehart and Winston Inc., New York, N. Y. (1967).

Feichtinger, H. G.: [1] (1973) Zur Ideal Theorie von Segal-Algebren, Manuscripta Math., 10 (1973), 307-312.

[2] Multipliers from $L^1(G)$ to a homogeneous Banach space, J. Math. Anal. Appl., to appear.

Figà-Talamanca, A. and Gaudry, G. T. (1970): Multipliers and sets of uniqueness of $L^p$, Michigan Math. J., 17 (1970), 179-191.

Friedberg, S. (1970): Closed subalgebras of group algebras, Trans. Amer. Math. Soc. 147 (1970), 117-125.

Goldberg, R. R.: [1] (1961) Fourier transforms, Cambrige University Press, Cambridge (1961).

[2] (1967) On a space of functions of Wiener, Duke Math. J., 34 (1967), 683-691.

[3] (1974) Recent results on Segal algebras, Springer-Verlag, Lecture Notes in Math., 399 (1974).

Helson, H. (1953): Isomorphism of abelian group algebras, Ark. Math., 2 (1953), 475-487.

Hewitt, E.: [1] (1958) A survey of abstract harmonic analysis. Some aspects of analysis and probability, 105-168. Surveys in Applied Math., 4 (1958), New York, N. Y.

[2] (1964) The ranges of certain convolution operators, Math. Scand., 15 (1964), 147-155.

Hewitt, E. & Kakutani, S.: [1] (1960) A class of multiplicative linear functions on the measure algebras of a locally compact abelian group, Ill. J. Math., 4 (1960), 553-574.

[2] (1964) Some multiplicative linear functions on M(G). Ann. of Math., 79 (1964), 489-505.

Hewitt, E. and Ross, K. A.: Abstract harmonic analysis, Springer Verlag, New York, I (1963), II (1970).

Iwasawa, K. (1944): On group rings of topological groups. Proc. Imp. Acad. Japan, Tokyo, 20 (1944), 67-70.

Johnson, B. E. (1973): Some examples in harmonic analysis. Studia Math., 48 (1973), 181-188.

Kahane, J. P.: [1] (1955) Sur les fonctions sommes de series trigonometriques absolument convergentes. C. R. Acad. Sci., Paris 240 (1955), 36-37.

[2] (1956) Sur certaines classes de series de Fourier absolument convergentes. J. Math. Pures Appl. 35 (1956) 249-259.

[3] (1965) Idempotents and closed subalgebras of A(Z). Proc. Internat. Symp. On Function Algebras, Tulane University (1965), 198-207.

Kaplansky, I. (1948): Dual rings. Ann. of Math. 49 (1948), 689-701.

Katznelson, Y. (1968): An introduction to harmonic analysis, John Wiley, New York, N. Y. (1968).

Krogstad, H. E.: [1] (1973) Multipliers on homogeneous Banach spaces on compact groups, Institut Mittag-Leffler, 1973.

[2] (1973) On Banach modules with weakly compact action, Institut Mittag-Leffler, 1973.

[3] (1974) Multipliers on Segal algebras, Department of Mathematics, University of Trondheim, Trondheim, Norway. See Burham J. T.

Lakien, E. H.: Nonfactorization in Segal algebras on compact abelian groups, Dissertation, Northwestern University.

Larsen, R.: [1] (1971) An introduction to the theory of multipliers, Springer-Verlag, 1971.

[2] (1973) Functional Analysis: An introduction, Dekber, New York, 1973.

[3] (1973) Banach Algebras: An introduction, Dekber, New York, 1973.

[4] (1973) Factorization and multipliers of Segal algebras, Preprint Series, Institute of Mathematics, University of Oslo, #5, 1973.

[5] (1974) Tensor product factorization and multipliers. Preprint Series, Institute of Math., University of Oslo, #13, 1974.

Larsen, R., Liu, T. S. and Wang, J. K.: [1] (1964) On the functions with Fourier transforms in $L_p$, Michigan Math. J., 11 (1964), 369-378.

See Burham J. T.

de Leeuw, K. (1958): Homogeneous algebras on compact abelian groups, Trans. Amer. Math. Soc., 87 (1958), 372-386.

Leibenson, Z. L. (1954): On the ring of functions with absolutely convergent Fourier series, Uspehi. Math. Nauk. N. S., 9 (1954), 157-162.

Leinert, M.: [1] (1973) A contribution to Segal algebras, Manuscripta Math., 10 (1973), 297-306.

[2] (1975) Remarks on Segal algebras, Manuscripta Math., 16 (1975), 1-9.

Lin, S. M. and Wang, H. C. (1972): Contributions to homogeneous Banach spaces. Special Issue Dediacted to the late Professor Puh Pan, The Chinese University of Hong Kong (1972), 151-120.

Liu, T. S., Rooij, A. van, and Wang, J. K. (1973): Projections and approximate identities for ideals in group algebras. Trans. Amer. Math. Soc. 175 (1973), 469-482.

Littlewood, J. E. (1944): Lectures on the theory of functions, London, Oxford University Press (1944).

Loomis, L. H. (1953): An introduction to abstract harmonic analysis, Van Nostrand Company, Inc-1 New York (1953).

Martin, J. C., and Yap, L. Y. H. (1970): The algebra of functions with Fourier transforms in $L^p$ . Proc. Amer. Math. Soc. 24 (1970), 217-219.

Paschke, W. L. (1973): A factorable Banach algebra without bounded approximate identity. Pacific J. Math. 46 (1973), 249-251.

Porcelli, P. (1966): Linear spaces of analytic functions, McGraw-hill, New York (1966).

Reiter,H.J.: [1] (1968) Classical harmonic analysis and locally compact groups, Oxford Mathematical Monographs, Oxford University Press, Oxford (1968).

[2] (1971) $L^1$-algebra and Segal algebras, Lecture Notes in Math. 231 (1971).

Richart, C. E. (1960): General theory of Banach algebras, D. Van Nostrand Co., Princeton, N. J. (1960).

Rider, D. (1969): Closed subalgebras of $L^1(T)$ . Duke Math. J. 36 (1969), 105-116.

Rudin, W.: [1] (1956) The automorphisms and the endomorphisms of the group algebra of the unit circle. Acta Math. 95 (1956).

[2] (1957) Factorization in the group algebra of the real line. Proc. Nat. Acad. Sci. U. S. A., 43 (1957), 339-340.

[3] (1958) Respresentation of functions by convolutions. J. Math. Mech. 7 (1958), 103-115.

[4] (1958) On isomorphisms of group algebras. Bull. Amer. Math. Soc. 64 (1958), 167-169.

[5] (1959) Measure algebras on abelian groups. Bull. Amer. Math. Soc. 65 (1959), 227-247.

[6] (1962) Fourier analysis on groups, Interscience, New York, N. Y. (1962).

Salem, R. (1945): Sur les transformations des séries de Fourier. Fund. Math. 33 (1945), 108-114.

Segal, I. E. (1947): The group algebra of a locally compact group. Trans. Amer. Math. Soc. 61 (1947), 69-105.

Sentilles, F. D., and Taylor, D. C. (1969): Factorization in Banach algebras and the general strict topology. Trans. Amer. Math. Soc. 142 (1969), 141-152.

Šilov, G. E.: Homogeneous rings of functions. Uspehi Matematičeskih Nauk (N.S.) 6, #1 (41)(1951), 91-137. Also Amer. Math. Soc. Translations, Series 1, 8 (1954), 393-455.

Taylor, D. C. (1968): A characterization of Banach algebras with approximate unit. Bull. Amer. Math. Soc. 74 (1968) 761-766.

Unni, K. R.: [1] (1972) Multipliers on a Segal algebras, Preprint, Matscience, Madras, 1972.

[2] (1972) Fourier transforms of the space of multipliers on a Segal algebra, Preprint, Matscience, Madras, 1972.

[3] (1972) Multipliers on the algebra $A_w^p(G)$ , Preprint, Matscience, Madras, 1972.

[4] (1972) Segal algebras of Beurling type, Preprint, Matscience, Madras, 1972.

[5] (1974) Parameasures and multipliers of Segal algebras. Lectures notes in Math., Springer-Verlag, 1974.

[6] (1974) A note on mutipliers of a Segal algebras, Studia Math., 49 (1974), 125-127.

Varopoulos, N. Th.: [1] (1964) Sur les forms positives d'une algebra de Banach, C. R. Acad. Sci. Paris, 258 (1964), 2465-2467.

[2] (1964) Continuite des forms lineaires positives sur une algebra de Banach avec involution., C. R. Acad. Sci. Paris, 258 (1964), 1121-1124.

[3] (1967) Tensor algebras and harmonic analysis, Acta Math., 119 (1967), 51-112.

Wang, H. C.: [1] (1972) Nonfactorization in group algebras, Studia Math., 42 (1972), 231-241.

[2] (1973) Factorization in homogeneous Banach algebras,

Chinese J. of Math. 1 (1973), 199-202.

[3] (1974) Some results on homogeneous Banach algebras, Tamkang J. of Math. 5 (1974), 149-159.

[4] Maximal homogeneous Banach spaces, Preprint.

Wang, H. C. and Lin, S. M. (1972): Contributions to Homogeneous Banach spaces, Special Issue Dedicated to the late Prof. Puh Pan, The Chinese University of Hong Kong, 1972, 115-120.

Wang, H. C. and Lee, Y. J.: [1] (1973) Homomorphisms of $A^p$-algebras, Chinese J. of Math. 1 (1973), 189-192.

[2] (1973) Complex homomorphisms of homogeneous Banach algebras, Tamkang J. of Math., 4 (1973), 29-33.

Wang, H. C. and Tseng C. N. (1975): Closed subalgebras of homogeneous Banach algebras, J. Austr. Math. Soc., 20 (1975), #3, 366-376.

Warner, C. R. (1966): Closed ideals in the group algebra $L^1(G) \cap L^2(G)$, Trans. Amer. Math. Soc., 121 (1966), 405-423.

Wichmann J. (1973): Bounded approximate units and bounded approximate identities, Proc. Amer. Math. Soc., 41 (1973), #2, 547-550.

Wiener, N. (1932): Tauberian theorems, Ann. of Math., (2) 33 (1932), 1-100.

Yap, L. Y. H.: [1] (1970) On the impossibility of representing certain functions by convolution, Math. Scand., 26 (1970), 132-140.

[2] (1970) Ideals in subalgebras of the group algebras, Studia Math., 35 (1970), 165-175.

[3] (1971) Every Segal algebra satisfies Ditkin's condition, Studia Math., 40 (1971), 235-237.

[4] Ideals in Fréchet subalgebras of $L^1(G)$ , Math. Balkanice, To appear.

[5] Nonfactorization of functions in Banach subspaces of $L^1(G)$ , Proc. Amer. Math. Soc., To appear.

Yap, L. Y. H. and Martin, J. C. (1970): The algebra of functions with Fourier transforms in $L^p$ , Proc. Amer. Math. Soc. 24 (1970), 217-219.

Zygmund, A. (1959): Trigonometric Series, 2nd ed., Cambridge University Press (1959).

# Index

C



D



E



F



G



H